LAGRANGIAN AND HAMILTONIAN MECHANICS

Solutions to the Exercises

LAGRANGIAN AND HAMILTONIAN MECHANICS

Solutions to the Exercises

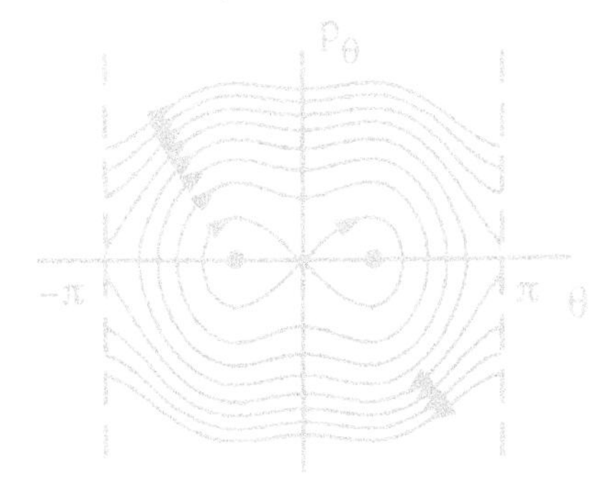

M. G. CALKIN

Dalhousie University, Canada

NEW JERSEY • LONDON • SINGAPORE • BEIJING • SHANGHAI • HONG KONG • TAIPEI • CHENNAI

Published by

World Scientific Publishing Co. Pte. Ltd.

5 Toh Tuck Link, Singapore 596224

USA office: 27 Warren Street, Suite 401-402, Hackensack, NJ 07601

UK office: 57 Shelton Street, Covent Garden, London WC2H 9HE

British Library Cataloguing-in-Publication Data
A catalogue record for this book is available from the British Library.

First published 1999
Reprinted 2006, 2008, 2011

LAGRANGIAN AND HAMILTONIAN MECHANICS: SOLUTIONS TO THE EXERCISES

ISBN-13 978-981-02-3782-0
ISBN-10 981-02-3782-0

PREFACE

This book contains the exercises from the intermediate/advanced classical mechanics text *Lagrangian and Hamiltonian Mechanics* (World Scientific Pub. Co. Pte. Ltd., Singapore, 1996) together with their complete solutions. In a few of the exercises I have seen fit to make minor changes in the wording; these are marked by asterisks. Also, I have not included the final Exercise 10-5, which is really an open-ended mini research project.

The present work is intended primarily for instructors who are using *Lagrangian and Hamiltonian Mechanics* in their course. It is hoped that it will assist them in choosing suitable assignments for their students, and that it will occasionally provide new insights. Instructors may also wish to photocopy and post the solutions to those exercises with which their students have had particular difficulty.

This book may also be used, together with *Lagrangian and Hamiltonian Mechanics*, by those who are studying mechanics on their own. In this case I strongly urge the individuals to make serious efforts to work out a substantial number of the relevant exercises on completing their study of each chapter and *before* looking at what I have written here. Only in this way will such individuals come face to face with, and hopefully overcome, the various difficulties which the exercises present. Exercises, whether mental or physical, are meant to be done, not read about!

Melvin G. Calkin
Halifax, Nova Scotia
September, 1998

CONTENTS

CHAPTER I

NEWTON'S LAWS

Exercise 1.01

A particle of mass m moves in one dimension x in a potential well

$$V = V_0 \tan^2(\pi x/2a)$$

where V_0 and a are constants. Find, for given total energy E, the position x as a function of time and the period τ of the motion. In particular, examine and interpret the low energy ($E << V_0$) and high energy ($E >> V_0$) limits of your expressions.

Solution

Energy conservation yields

$$\tfrac{1}{2}m\dot{x}^2 + V_0 \tan^2(\pi x/2a) = E.$$

This can be rearranged to give

$$t = \sqrt{\frac{m}{2}} \int_{x_0}^{x} \frac{dx}{\sqrt{E - V_0 \tan^2(\pi x/2a)}} = \sqrt{\frac{m}{2}} \int_{x_0}^{x} \frac{\cos(\pi x/2a)\, dx}{\sqrt{E - (E + V_0)\sin^2(\pi x/2a)}}.$$

To do the x-integration, we set

$$\sin\left(\frac{\pi x}{2a}\right) = \sqrt{\frac{E}{E + V_0}} \sin\phi, \qquad \frac{\pi}{2a}\cos\left(\frac{\pi x}{2a}\right) dx = \sqrt{\frac{E}{E + V_0}} \cos\phi\, d\phi,$$

and hence obtain

$$\phi - \phi_0 = \frac{\pi}{2a}\sqrt{\frac{2(E + V_0)}{m}}\, t.$$

The motion is thus given by (Fig. 1)

$$\sin\left(\frac{\pi x}{2a}\right) = \sqrt{\frac{E}{E + V_0}} \sin\left(\frac{\pi}{2a}\sqrt{\frac{2(E + V_0)}{m}}\, t + \phi_0\right).$$

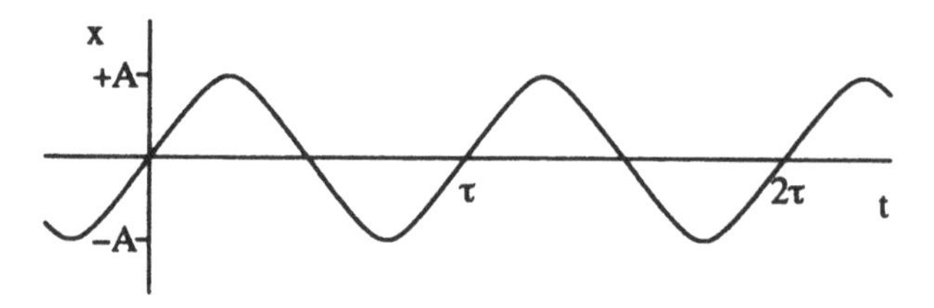

Ex. 1.01, Fig. 1

The motion is periodic with period

$$\tau = 4a\sqrt{\frac{m}{2(E+V_0)}}.$$

The turning points $\pm A$ of the motion are given by

$$\sin\left(\frac{\pi A}{2a}\right) = \sqrt{\frac{E}{E+V_0}} \qquad \left(\text{or by } \tan\left(\frac{\pi A}{2a}\right) = \sqrt{\frac{E}{V_0}}\right).$$

For $E << V_0$ we have $\pi A/2a << 1$ and can use the small angle approximation to write

$$\frac{\pi x}{2a} \approx \sqrt{\frac{E}{V_0}}\sin\left(\frac{\pi}{2a}\sqrt{\frac{2V_0}{m}}\,t + \phi_0\right).$$

This is simple harmonic motion with angular frequency $\omega_0 = \frac{\pi}{2a}\sqrt{\frac{2V_0}{m}}$ and amplitude $\frac{2a}{\pi}\sqrt{\frac{E}{V_0}} = \sqrt{\frac{2E}{m\omega_0^2}}$. This is to be expected, since for small amplitude motion the potential approximates the "harmonic oscillator potential"

$$V_0(\pi x/2a)^2 = \tfrac{1}{2}m\omega_0^2 x^2.$$

For $E >> V_0$ we have

$$\sin\left(\frac{\pi x}{2a}\right) \approx \sin\left(\frac{\pi}{2a}\sqrt{\frac{2E}{m}}\,t + \phi_0\right)$$

so, now dropping the arbitrary phase,

$$\frac{\pi x}{2a} \approx \frac{\pi}{2a}\sqrt{\frac{2E}{m}}\,t - 2n_+\pi \quad \text{or} \quad -\frac{\pi}{2a}\sqrt{\frac{2E}{m}}\,t + (2n_- + 1)\pi \qquad \text{with} \qquad n_\pm = 0, \pm 1, \pm 2, \cdots.$$

That is,

$$x \approx \sqrt{\frac{2E}{m}}\,t - 4an_{+} \quad \text{or} \quad -\sqrt{\frac{2E}{m}}\,t + 4a(n_{-} + \tfrac{1}{2}).$$

The integers $n_{\pm}$ must be chosen appropriately; see Fig. 2.

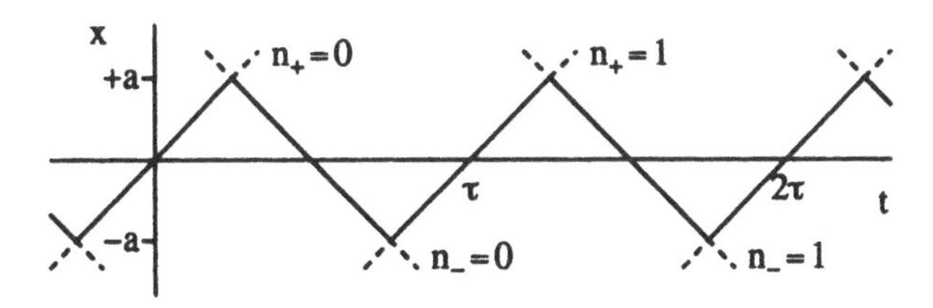

Ex. 1.01, Fig. 2

In this limit the particle oscillates back and forth at constant speed $\sqrt{2E/m}$ between rigid walls at $x = \pm a$. This is to be expected, since in this limit the potential approximates the "infinite square well potential"

$$\begin{aligned} V &= 0 \quad \text{for } |x| < a \\ V &\to \infty \quad \text{for } |x| > a\,. \end{aligned}$$

Exercise 1.02

For each of the following central potentials $V(r)$ sketch the effective potential

$$V_{eff}(r) = \frac{L^2}{2mr^2} + V(r),$$

and use your sketch to classify and draw qualitative pictures of the possible orbits.

(a) $V(r) = \frac{1}{2}kr^2$ **3D isotropic harmonic oscillator**

(b) $V(r) = -V_1$ for $r < a$ **square well**
$V(r) = 0$ for $r > a$

(c) $V(r) = -\dfrac{k}{r^2}$

(d) $V(r) = -\dfrac{k}{r^4}$

(e) $V(r) = -k\dfrac{e^{-\alpha r}}{r}$ **Yukawa potential**

Note that the qualitative shape of $V_{eff}(r)$ versus r may depend on L and on the various parameters; consider all cases (but assume that the given parameters are positive).

Solution

(a) The effective potential is

$$V_{eff}(r) = \frac{L^2}{2mr^2} + \frac{1}{2}kr^2.$$

For $r \to 0$ the term $L^2/2mr^2$ dominates and $V_{eff} \to \infty$. For $r \to \infty$ the term $\frac{1}{2}kr^2$ dominates and again $V_{eff} \to \infty$. The effective potential has a minimum at

$$\frac{dV_{eff}}{dr} = -\frac{L^2}{mr^3} + kr = 0;$$

that is, at $r = r_0$ where $r_0^4 = L^2/mk$. The minimum of the effective potential is $V_{eff}(r_0) = L\sqrt{k/m} = E_0$. This effective potential is shown in Fig. 1.

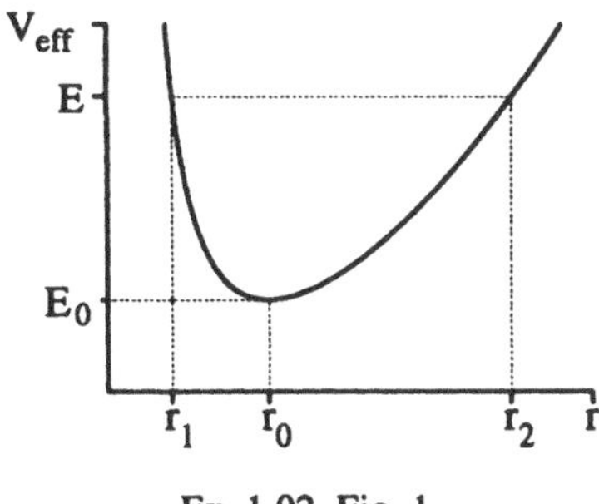

Ex. 1.02, Fig. 1

For $E = E_0$ the orbit radius is fixed at r_0 and the orbit is a circle. For $E > E_0$ the radius oscillates back and forth between turning radii r_1 and r_2 and the orbit looks qualitatively like Fig. 2.

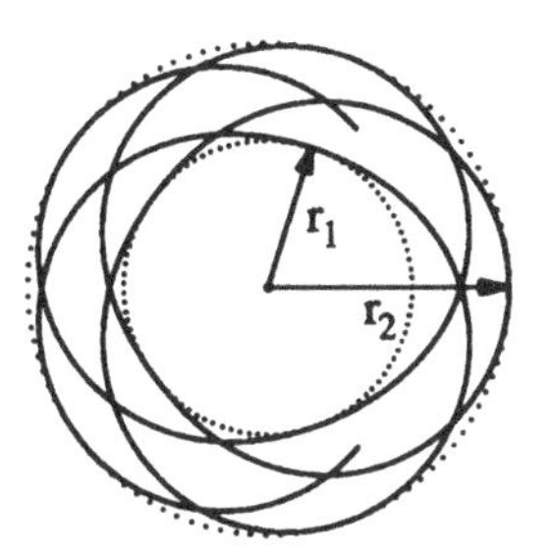

Ex. 1.02, Fig. 2

The motion of a particle in this potential is studied in detail in Exercise 1.14. It turns out that the orbit is actually an ellipse with geometric center at the force center (Fig. 3).

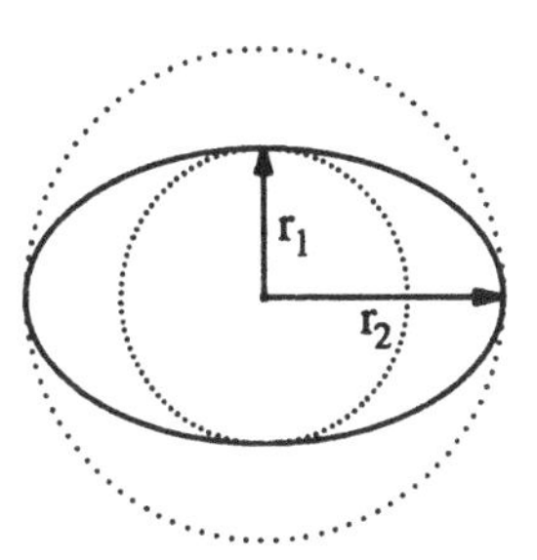

Ex. 1.02, Fig. 3

(b) The effective potential is

$$V_{eff} = \begin{cases} L^2/2mr^2 - V_1 & \text{for } r < a \\ L^2/2mr^2 & \text{for } r > a \end{cases}$$

For $r \to 0$ the term $L^2/2mr^2$ dominates and $V_{eff} \to \infty$. For $r \to \infty$ $V_{eff} \to 0$ from above. Further, V_{eff} has a discontinuity of V_1 at $r = a$. There are two possibilities for V_{eff}, depending on the magnitude of V_1. These are shown in Fig. 4. For $V_1 > L^2/2ma^2$ the effective potential goes negative (Fig. 4(a)) whereas for $V_1 < L^2/2ma^2$ the effective potential is always positive (Fig. 4(b)).

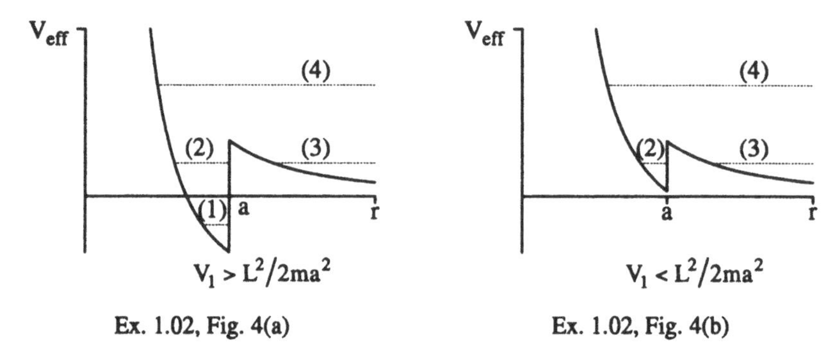

Ex. 1.02, Fig. 4(a) Ex. 1.02, Fig. 4(b)

For situation (1) or (2) in Fig. (4) the orbit radius oscillates back and forth between an inner turning radius r_1 and an outer turning radius $r_2 = a$. Although the orbits for (1) and (2) are similar it is useful to distinguish them, (1) having negative and (2) having positive energy relative to infinity. The orbit looks qualitatively like Fig. 2. In fact, since the force on the particle is zero for $r < a$, the orbit is composed of straight line segments, Fig. 5.

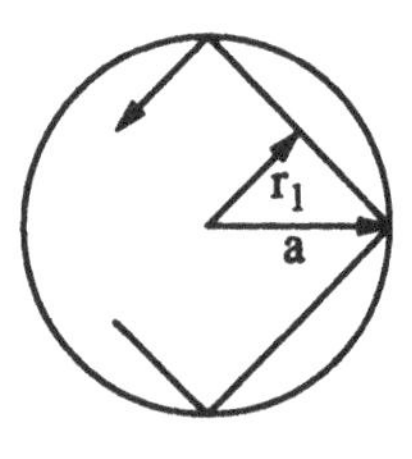

Ex. 1.02, Fig. 5

For situation (3) in Fig. (4) the orbit radius decreases from infinity to a minimum radius $r_3 > a$ and then increases back to infinity. The orbit looks qualitatively like Fig. 6.

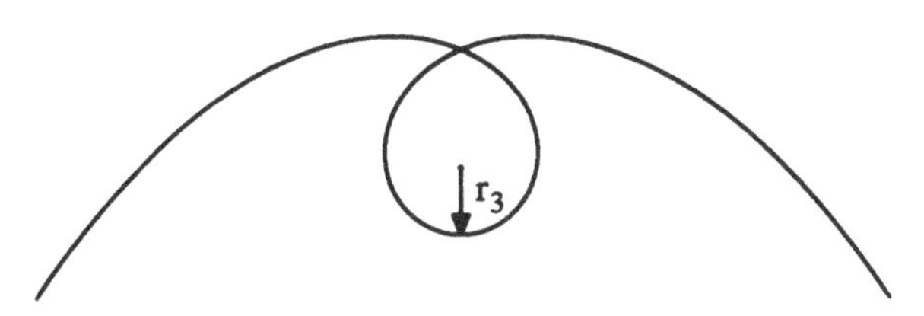

Ex. 1.02, Fig. 6

In fact, since in this case the particle never encounters a force, the orbit is a straight line which passes by the potential region, Fig. 7.

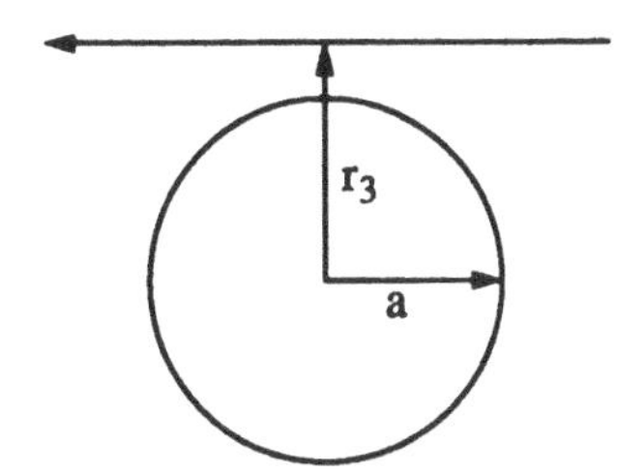

Ex. 1.02, Fig. 7

For situation (4) in Fig. (4) the orbit radius decreases from infinity to a minimum radius $r_3 < a$ and then increases back to infinity. The orbit again looks qualitatively like Fig. 6. In fact, since in this case the particle encounters a force only at $r = a$ the orbit is composed of straight line segments, Fig. 8.

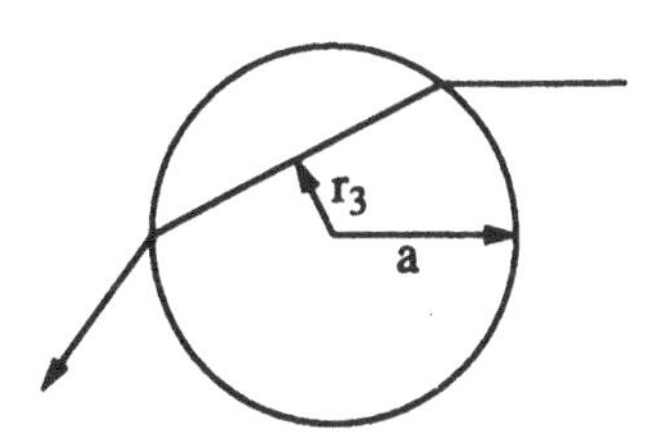

Ex. 1.02, Fig. 8

(c) The effective potential is

$$V_{\text{eff}}(r) = \left(\frac{L^2}{2mk} - 1\right)\frac{k}{r^2}.$$

The shape of V_{eff} depends on the magnitude of the dimensionless combination of parameters, $L^2/2mk$. For $L^2/2mk < 1$, $V_{eff} \propto -1/r^2$ (Fig. 9(a)) whereas for $L^2/2mk > 1$, $V_{eff} \propto +1/r^2$ (Fig. 9(b)).

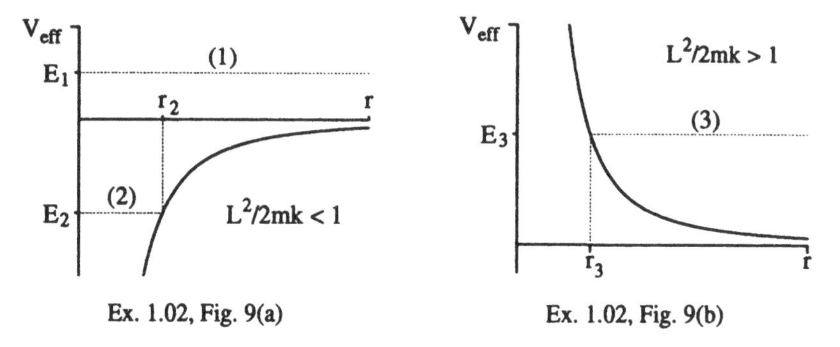

Ex. 1.02, Fig. 9(a) Ex. 1.02, Fig. 9(b)

For $E > 0$ and $L^2/2mk < 1$ (situation (1)) there is no inner turning radius and the particle spirals in to the force center. The orbit is a capture orbit like Fig. 10.

Ex. 1.02, Fig. 10

Indeed, since $L = b\sqrt{2mE}$ where b is the impact parameter, we can write the condition for capture as $b^2 < k/E$. This gives a capture cross section $\sigma_{capture} = \pi b_{max}^2 = \pi k/E$. For $E < 0$ and $L^2/2mk < 1$ (situation (2)) the orbit has an outer turning radius r_2 but no inner turning radius and the orbit again spirals in to $r = 0$. For $E > 0$ and $L^2/2mk > 1$ (situation (3)), the orbit is a scattering orbit like Fig. 6.

(d) The effective potential is

$$V_{eff}(r) = \frac{L^2}{2mr^2} - \frac{k}{r^4}.$$

For $r \to 0$ the term $-k/r^4$ dominates and $V_{eff} \to -\infty$. For $r \to \infty$ the term $L^2/2mr^2$ dominates and $V_{eff} \to 0$ from above. There is one axis crossing at

$$V_{eff} = \frac{L^2}{2mr^2} - \frac{k}{r^4} = 0;$$

that is, at $r^2 = 2mk/L^2$. There is one extremum (a maximum) at

$$\frac{dV_{eff}}{dr} = -\frac{L^2}{mr^3} + \frac{4k}{r^5} = 0;$$

that is, at $r^2 = 4mk/L^2$. The value of the effective potential at its extremum is $V_{eff}(\max) = L^4/16m^2k = b^4E^2/4k$ where b is the impact parameter. The resulting effective potential is shown in Fig. 11.

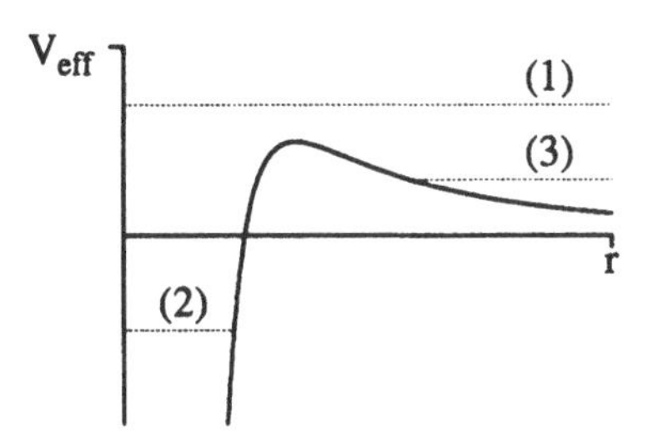

Ex. 1.02, Fig. 11

For $E > V_{eff}(\max)$ and thus $b^4 < 4k/E$ (situation (1)) the orbit is a capture orbit like Fig. 10. The capture cross section is $\sigma_{capture} = \pi b_{max}^2 = 2\pi\sqrt{k/E}$. For $E < V_{eff}(\max)$ and initial orbit radius less than $\sqrt{4mk/L^2}$ (situation (2)) the orbit has an outer turning radius but again no inner turning radius and the orbit spirals in to the force center. For $E < V_{eff}(\max)$ and initial radius greater than $\sqrt{4mk/L^2}$ (situation (3)) the orbit is a scattering orbit like Fig. 6.

(e) The effective potential is

$$V_{eff}(r) = \frac{L^2}{2mr^2} - k\frac{e^{-\alpha r}}{r}$$

For $r \to 0$ the term $L^2/2mr^2$ dominates and $V_{eff} \to \infty$. For $r \to \infty$ the term $L^2/2mr^2$ again dominates (the exponential causes the second term to decrease faster than any negative power of r, as $r \to \infty$) and $V_{eff} \to 0$ *from above.*

To determine the shape of V_{eff} in between, we first check whether or not there are any axis crossings (in view of the limiting behavior there must be zero or an even number of them). These occur at radii for which

$$\alpha L^2/2mk = \alpha r e^{-\alpha r}.$$

The right-hand side of this equation, as a function of αr, is shown in Fig. 12.

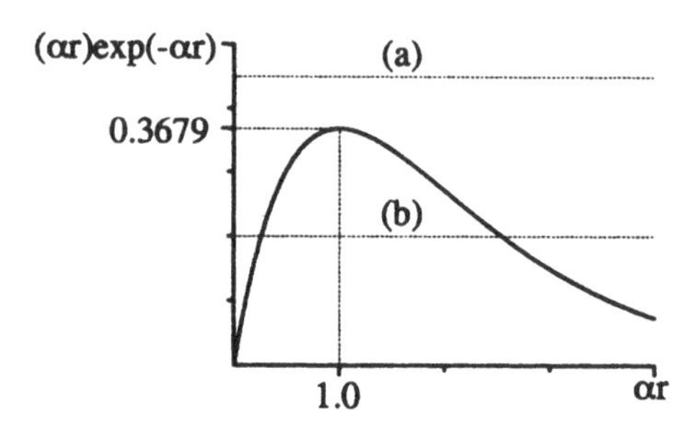

Ex. 1.02, Fig. 12

The peak in Fig. (12) is at $\alpha r = 1$ and the height of the peak is $1/e = 0.3679$. Thus, if the dimensionless combination of parameters $\alpha L^2/2mk$ is greater than 0.3679, there are no axis crossings (situation (a)), and if $\alpha L^2/2mk$ is less than 0.3679, there are two axis crossings (situation (b)).

We next check whether or not there are any extrema. These occur at radii for which

$$\alpha L^2/2mk = (\alpha r/2)(1+\alpha r)e^{-\alpha r}.$$

The right-hand side of this equation, as a function of αr, is shown in Fig. 13.

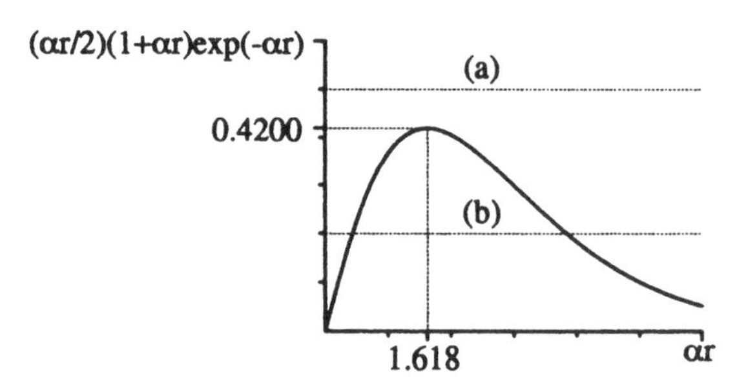

Ex. 1.02, Fig. 13

The peak in Fig. (13) is at $\alpha r = (1+\sqrt{5})/2 = 1.618$ and the height of the peak is 0.4200. Thus, if $\alpha L^2/2mk$ is greater than 0.4200, there are no extrema (situation (a)), and if $\alpha L^2/2mk$ is less than 0.4200, there are two extrema (situation (b)).

The possibilities for V_{eff} are shown in Fig. 14.

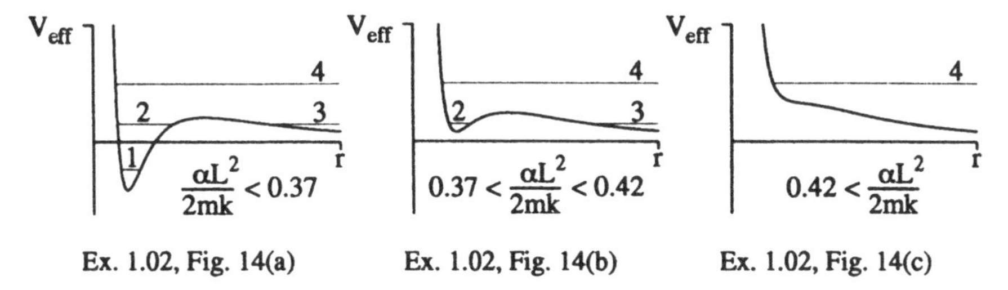

Ex. 1.02, Fig. 14(a) Ex. 1.02, Fig. 14(b) Ex. 1.02, Fig. 14(c)

Situations (1) and (2) in Fig. 14 give bound orbits as in Ex. 1.02, Fig. 2, and situations (3) and (4) give scattering orbits as in Ex. 1.02, Fig. 6.

Exercise 1.03

The first U.S. satellite to go into orbit, Explorer I, which was launched on January 31, 1958, had a perigee of 360 km and an apogee of 2549 km above the earth's surface. Find:
(a) the semi-major axis,
(b) the eccentricity,
(c) the period,
of Explorer I 's orbit. The earth's equatorial radius is 6378 km and the acceleration due to gravity at the earth's surface is $g = 9.81\,\mathrm{m/s^2}$.

Solution

The minimum and maximum radii of Explorer I's orbit are

$$a(1-e) = 360 + 6378 = 6738\,\mathrm{km} \quad \text{and} \quad a(1+e) = 2549 + 6378 = 8927\,\mathrm{km}.$$

The semi-major axis is thus $a = \frac{1}{2}(6738 + 8927) = 7833\mathrm{km}$ and the eccentricity is $e = \dfrac{8927 - 6738}{8927 + 6738} = 0.14$. The period of the orbit is given by

$$\tau = \frac{2\pi}{\sqrt{GM}} a^{3/2}.$$

Rather than looking up the gravitational constant G and the mass M of the earth in a handbook, it is simpler to observe that the gravitational field at the surface of the earth is $g = GM/R^2 = 9.81\,\mathrm{m/s^2}$, and thus

$$\tau = \frac{2\pi}{\sqrt{9.81 \times (6.378 \times 10^6)^2}} (7.833 \times 10^6)^{3/2} = 6895\,\mathrm{s} = 1.92\,\mathrm{hr}.$$

Exercise 1.04

Mars travels on an approximately elliptical orbit around the Sun. Its minimum distance from the Sun is about 1.38 AU and its maximum distance is about 1.67 AU (1 AU = mean distance from Earth to Sun). Find:
(a) the semi-major axis,
(b) the eccentricity,
(c) the period,
of Mars' orbit.

Solution

The semi-major axis of Mars' orbit is

$$a = \tfrac{1}{2}(1.38 + 1.67) = 1.53\,\text{AU},$$

and the eccentricity is

$$e = \frac{1.67 - 1.38}{1.67 + 1.38} = 0.095.$$

The period is given by

$$\tau = \frac{2\pi}{\sqrt{GM}} a^{3/2}.$$

Rather than looking up the gravitational constant G and the mass M of the Sun, it is simpler to observe that for Earth

$$1\,\text{year} = \frac{2\pi}{\sqrt{GM}}(1\,\text{AU})^{3/2},$$

and thus for Mars

$$\tau,\ \text{in years} = (a,\ \text{in AU})^{3/2} = (1.53)^{3/2} = 1.88.$$

Exercise 1.05

The most economical method of traveling from one planet to another, the Hohmann transfer, consists of moving along a (Sun-controlled) elliptical path which is tangent to the (approximately) circular orbits of the two planets. Consider a Hohmann transfer from Earth (orbit radius 1.00 AU) to Venus (orbit radius 0.72 AU). Find, in units of AU and year:
(a) the semi-major axis of the transfer orbit,
(b) the time required to go from Earth to Venus,
(c) the velocity "kick" needed to place a spacecraft in Earth orbit into the transfer orbit.

In this problem ignore the effects of the gravitational fields of Earth and Venus on the spacecraft.

Solution

(a) The maximum radius of the transfer orbit equals the radius of Earth's orbit (1.00 AU), and the minimum radius equals the radius of Venus' orbit (0.72 AU). The semi-major axis of the transfer orbit is thus

$$a = \tfrac{1}{2}(1.00 + 0.72) = 0.86 \text{ AU}.$$

(b) The period τ of the transfer orbit is (see Exercise 1.04)

$$\tau = a^{3/2} = (0.86)^{3/2} = 0.80 \text{ year}.$$

The time to go from Earth to Venus is half the period, $\tfrac{1}{2}(0.80) = 0.40$ year.

(c) The energy of the spacecraft, per unit mass, at Earth orbit radius r_E is given by

$$\frac{1}{2}v^2 - \frac{GM}{r_E} = -\frac{GM}{2a}$$

where v is the speed of the spacecraft. On the right we have expressed the total energy of the spacecraft, per unit mass, in terms of the semi-major axis a of the transfer orbit. This equation gives

$$v^2 = GM\left(\frac{2}{r_E} - \frac{1}{a}\right).$$

Newton's second law shows that the speed v_E of the Earth around the Sun is given by

$$v_E^2 = \frac{GM}{r_E}.$$

Thus

$$\left(\frac{v}{v_E}\right)^2 = 2 - \frac{r_E}{a} = 2 - \frac{1}{0.86},$$

so $v/v_E = 0.915$. Hence the velocity "kick" $\Delta v = v - v_E$ needed to place the spacecraft into the transfer orbit is given by $\Delta v/v_E = -0.085$. Since the speed of the Earth is $v_E = 2\pi$ AU/year, this yields $\Delta v = -0.534$ AU/year ($= -2.53$ km/s).

Exercise 1.06

Halley's comet travels around the Sun on an approximately elliptical orbit of eccentricity $e = 0.967$ and period 76 years. Find:
(a) the semi-major axis of the orbit (Ans. 17.9 AU),
(b) the distance of closest approach of Halley's comet to the Sun (Ans. 0.59 AU),
(c) the time per orbit that Halley's comet spends within 1 AU of the Sun (Ans. 78 days).

Solution

(a) The semi-major axis of the orbit of Halley's comet is given by Kepler's third law (see Exercise 1.04), $a = \tau^{2/3} = (76)^{2/3} = 17.9$ AU.
(b) The minimum distance of Halley's comet to the Sun is $a(1-e) = 0.59$ AU.

(c) The instantaneous distance r of Halley's comet to the Sun and the eccentric anomaly ψ are related by

$$\frac{r}{a} = 1 - e\cos\psi .$$

In particular, the eccentric anomaly corresponding to a distance of 1 AU is given by

$$\frac{1}{17.9} = 1 - 0.967\cos\psi ,$$

so $\psi = 0.218$ radians. The corresponding time is then given by Kepler's equation

$$\psi - e\sin\psi = (2\pi/\tau)t$$

which becomes

$$0.218 - 0.967\sin 0.218 = (2\pi/76)t ,$$

so $t = 0.107$ year $= 39$ day. The total time per orbit that Halley's comet spends within 1 AU of the Sun is $2 \times 39 = 78$ days.

Exercise 1.07

Define a "season" as a time interval over which the true anomaly increases by $\pi/2$. Find the duration of the shortest season for earth. Take the eccentricity of earth's orbit to be 0.0167.

Solution

The true anomaly θ for the earth, measured from perihelion, and the eccentric anomaly ψ are related by

$$\left(\frac{r}{a}=\right) \quad 1-e\cos\psi=\frac{1-e^2}{1+e\cos\theta}$$

where $e = 0.0167$ is the eccentricity of earth's orbit. This equation can be written

$$\cos\psi=\frac{e+\cos\theta}{1+e\cos\theta}.$$

For the shortest "season" the true anomaly increases from $-\pi/4$ to $\pi/4$, and the eccentric anomaly increases from $-\psi$ to ψ where

$$\cos\psi=\frac{0.0167+\cos(\pi/4)}{1+0.0167\cos(\pi/4)}=0.7154.$$

This yields $\psi = 0.7737$ radians. Kepler's equation (with t in years),

$$2\pi t=\psi-e\sin\psi=0.7737-0.0167\sin 0.7737=0.7620,$$

then shows that t increases from -0.1213 to 0.1213, so the duration of the shortest "season" is $2\times 0.1213=0.2425$ year $=88.59$ days. Compare this with "winter" ($=89.00$ days).

Exercise 1.08

A satellite of mass m moves in a circular orbit of radius a_0 around the earth.
(a) A rocket on the satellite fires a burst radially, and as a result the satellite acquires, essentially instantaneously, a radial velocity u in addition to its angular velocity. Find the semi-major axis, the eccentricity, and the orientation of the elliptical orbit into which the satellite is thrown.
(b) Repeat (a), if instead the rocket fires a burst tangentially.
(c) In both cases find the velocity kick required to throw the satellite into a parabolic orbit.

Solution

Newton's second law shows that

$$\frac{mv_0^2}{a_0}=\frac{k}{a_0^2},$$

so the initial speed of the satellite is $v_0=\sqrt{k/ma_0}$. The initial energy is

$$E_0=\frac{1}{2}mv_0^2-\frac{k}{a_0}=-\frac{1}{2}mv_0^2=-\frac{k}{2a_0},$$

and the initial angular momentum is

$$L_0 = mv_0a_0 = \sqrt{mka_0}\,.$$

(a) A radial kick changes the energy of the satellite. Since the velocities v_0 and u are perpendicular to one another, the final energy is

$$E = E_0 + \frac{1}{2}mu^2.$$

Expressing the initial and final energies in terms of the semi-major axes of the orbits, we have

$$-\frac{k}{2a} = -\frac{k}{2a_0} + \frac{1}{2}mu^2.$$

This can be rearranged to give the relation between the initial and final semi-major axes,

$$\frac{a_0}{a} = 1 - \frac{u^2}{v_0^2}.$$

A radial kick does not change the angular momentum of the satellite, so

$$\frac{L^2}{mk} = \frac{L_0^2}{mk}.$$

Expressing this in terms of the semi-major axes and eccentricities of the orbits, we have

$$a(1 - e^2) = a_0.$$

Substituting the preceding expression for a_0/a into this equation, we obtain the eccentricity of the final orbit

$$e = |u|/v_0\,.$$

The equation for the final orbit is

$$\frac{a_0}{r} = 1 + e\cos\theta$$

where θ is the angle from pericenter. At the kick $r = a_0$ and, for u positive, is increasing. The kick thus occurs at $\theta = +\pi/2$. That is, pericenter is at $-\pi/2$ from the point of firing of the rocket.

For $|u| \to v_0$ the eccentricity $e \to 1$ and the semi-major axis $a \to \infty$ such that $a(1-e) \to a_0/2$. The satellite is then thrown into a parabolic orbit. The energy put into the system is $\frac{1}{2}mv_0^2$, so the final total energy is zero, as required.

(b) A tangential kick changes the energy of the satellite. The final energy is

$$E = \frac{1}{2}m(v_0 + u)^2 - \frac{k}{a_0} = E_0 + mv_0u + \frac{1}{2}mu^2.$$

Expressing the energies in terms of the semi-major axes, we have

$$-\frac{k}{2a} = -\frac{k}{2a_0} + mv_0u + \frac{1}{2}mu^2.$$

This can be rearranged to give the relation between the initial and final semi-major axes,

$$\frac{a_0}{a} = 1 - 2\frac{u}{v_0} - \frac{u^2}{v_0^2}.$$

A tangential kick also changes the angular momentum of the satellite. The final angular momentum is

$$L = m(v_0 + u)a_0 = L_0(1 + u/v_0),$$

and thus

$$\frac{L^2}{mk} = \frac{L_0^2}{mk}\left(1 + \frac{u}{v_0}\right)^2.$$

Expressing this in terms of the semi-major axes and eccentricities of the orbits, we have

$$a(1 - e^2) = a_0(1 + u/v_0)^2.$$

Substituting the preceding expression for a_0/a into this equation, we obtain the eccentricity of the final orbit

$$e = \left|\left(\frac{u}{v_0}\right)\left(2 + \frac{u}{v_0}\right)\right|.$$

For $0 < u/v_0 < \sqrt{2} - 1$ or for $-\sqrt{2} - 1 < u/v_0 < -2$, $e = (u/v_0)(2 + u/v_0)$ and $a(1 - e) = a_0$. The equation for the final orbit is then

$$\frac{a_0(1+e)}{r} = 1 + e\cos\theta.$$

At the kick $r = a_0$ and thus $\theta = 0$. That is, pericenter is at the point of firing of the rocket. For $-2 < u/v_0 < 0$, $e = -(u/v_0)(2 + u/v_0)$ and $a(1 + e) = a_0$. The equation for the final orbit is then

$$\frac{a_0(1-e)}{r} = 1 + e\cos\theta.$$

At the kick $r = a_0$ and thus $\theta = \pi$. That is, apocenter is at the point of firing of the rocket.

For $u/v_0 \to \pm\sqrt{2} - 1$ (the final velocity is then $v_0 + u = \pm\sqrt{2}\,v_0$) the eccentricity $e \to 1$ and the semi-major axis $a \to \infty$. The satellite is then thrown into a parabolic orbit. The energy put into the system is $\frac{1}{2}m(\pm\sqrt{2}v_0)^2 - \frac{1}{2}mv_0^2 = \frac{1}{2}mv_0^2$, so the final total energy is zero, as required.

It is also of interest to consider the case $u/v_0 = -1$ (for which the velocity of the satellite immediately after rocket firing is zero). We then have $a = a_0/2$ and $e = 1$, and the ellipse degenerates into a straight line, the satellite falling straight in towards the center of the earth.

Exercise 1.09

Show that the following ancient picture of planetary motion (in heliocentric terms) is in accord with Kepler's picture, if the eccentricity e is small and terms of order e^2 and higher are neglected:

(a) the earth moves around the sun in a circular orbit of radius a; however, the sun is not at the center of this circle, but is displaced from the center by a distance ea;

(b) the earth does not move uniformly around the circle; however, a radius vector from a point which is on a line from the sun to the center, the same distance from and on the opposite side of the center as the sun, to the earth does rotate uniformly.

Solution

(a)

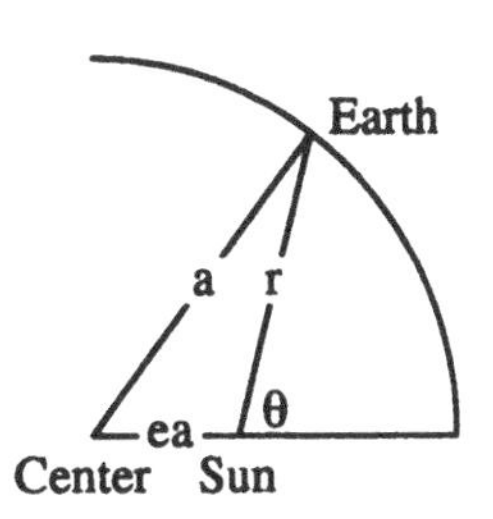

Ex. 1.09, Fig. 1

For an off-center circular orbit (Fig. 1) we have

$$a^2 = r^2 + e^2a^2 + 2ear\cos\theta,$$

as follows from the trigonometric cosine law. This yields

$$\frac{r}{a} = \sqrt{1 - e^2\sin^2\theta} - e\cos\theta = 1 - e\cos\theta - \tfrac{1}{2}e^2\sin^2\theta + \cdots.$$

On the other hand, for an elliptic orbit we have (*Lagrangian and Hamiltonian Mechanics*, pages 12 and 13)

$$\frac{r}{a} = \frac{1 - e^2}{1 + e\cos\theta} = 1 - e\cos\theta - e^2\sin^2\theta + \cdots.$$

The two expansions agree to order e.

(b)

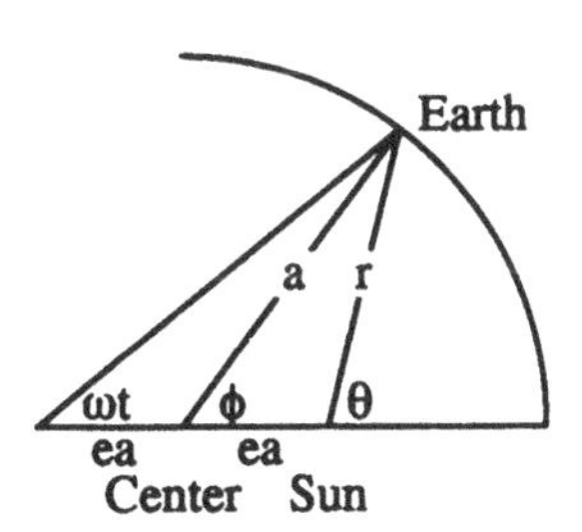

Ex. 1.09, Fig. 2

Applying the trigonometric sine law to the two smaller triangles in Fig. 2, we have

$$\sin(\phi - \omega t) = e\sin\omega t \quad \text{and} \quad \sin(\theta - \phi) = e\sin\theta.$$

These yield, to order e^2,

$$\phi = \omega t + e\sin\omega t + \cdots \quad \text{and} \quad \theta = \phi + e\sin\phi + \tfrac{1}{2}e^2\sin 2\phi + \cdots,$$

so for the "ancient picture" of planetary orbits we have

$$\theta = \omega t + 2e\sin\omega t + e^2\sin 2\omega t + \cdots.$$

On the other hand, for a Keplerian orbit we have (*Lagrangian and Hamiltonian Mechanics*, page 18)

$$\theta - 2e\sin\theta + \tfrac{3}{4}e^2\sin 2\theta + \cdots = \omega t.$$

Inversion gives

$$\theta = \omega t + 2e\sin\omega t + \tfrac{5}{4}e^2\sin 2\omega t + \cdots,$$

which agrees with the "ancient picture" to order e.

Exercise 1.10

(a) Show that

$$\frac{2r_0}{r} = 1 + \cos\theta$$

(the standard form for a conic section, on setting the eccentricity $e = 1$ and the semi-latus-rectum $p = 2r_0$) is the equation of a parabola, by translating it into cartesian coordinates with the origin at the focus and the x-axis through pericenter.
(b) A comet travels around the Sun on a parabolic orbit. Show that the distance r of the comet from the Sun is related to the time t from perihelion by

$$\frac{\sqrt{2}}{3}(r + 2r_0)\sqrt{r - r_0} = 2\pi t$$

where distances are measured in AU and time is measured in years.
(c) If one approximates the orbit of Halley's comet near the Sun by a parabola with $r_0 = 0.59\,\text{AU}$, what does this give for the time Halley's comet spends within 1 AU of the Sun?
(d) What is the maximum time a comet on a parabolic orbit may spend within 1 AU of the Sun?

Solution

(a) The equation for the orbit,

$$\frac{2r_0}{r} = 1 + \cos\theta,$$

can be transformed into cartesian coordinates (x, y) by multiplying it through by r and setting $r = \sqrt{x^2 + y^2}$ and $r\cos\theta = x$. We obtain

$$2r_0 - x = \sqrt{x^2 + y^2}.$$

Squaring and simplifying this, we then find

$$x = r_0 - \frac{y^2}{4r_0}$$

which is indeed the equation of a parabola in cartesian coordinates.

(b) To find out how r depends on time, we turn to the energy equation

$$\frac{1}{2}m\dot{r}^2 + \frac{L^2}{2mr^2} - \frac{k}{r} = E$$

and set $E = 0$ and $L^2 = 2mkr_0$ to obtain

$$\frac{1}{2}m\dot{r}^2 + \frac{kr_0}{r^2} - \frac{k}{r} = 0.$$

Multiplying by r^2, rearranging, and integrating then gives

$$\sqrt{\frac{2k}{m}}\,t = \int_{r_0}^{r} \frac{r\,dr}{\sqrt{r - r_0}} = \int_{r_0}^{r} \left[\sqrt{r - r_0} + \frac{r_0}{\sqrt{r - r_0}}\right] dr = \frac{2}{3}(r + 2r_0)\sqrt{r - r_0}.$$

Since $\sqrt{k/m} = \sqrt{GM} = 2\pi\,\mathrm{AU}^{3/2}/\mathrm{year}$, this becomes

$$2\pi t = \frac{\sqrt{2}}{3}(r + 2r_0)\sqrt{r - r_0}$$

with distances now in AU and time in years.

(c) For $r_0 = 0.59\,\text{AU}$ and $r = 1.00\,\text{AU}$ the preceding equation gives $t = 0.1047\,\text{year}$, so the time, in the parabolic approximation, that Halley's comet spends within 1 AU of the sun is $2t = 0.2094\,\text{year} = 76.5\,\text{days}$. Compare this with the actual time of 78 days (see Exercise 1.06).

(d) We set $r = 1.00$ and adjust r_0 to give maximum time t. This occurs for

$$2\pi\frac{dt}{dr_0} = \frac{\sqrt{2}}{3}\left[2\sqrt{1-r_0} - \frac{1+2r_0}{2\sqrt{1-r_0}}\right] = \frac{\sqrt{2}}{3}\frac{3-6r_0}{2\sqrt{1-r_0}} = 0;$$

that is, for $r_0 = 1/2$. The corresponding t is given by

$$2\pi t = \frac{\sqrt{2}}{3}\left(1 + 2\times\frac{1}{2}\right)\sqrt{1-\frac{1}{2}} = \frac{2}{3},$$

so $t = (1/3\pi)\,\text{year} = 38.75\,\text{days}$. The maximum time a comet on a parabolic orbit may spend within 1 AU of the sun is thus $2t = 77.5\,\text{days}$.

Exercise 1.11

A particle of mass m moves in a central force field $\mathbf{F} = -(k/r^2)\hat{\mathbf{r}}$.
(a) By integrating Newton's second law $d\mathbf{p}/dt = \mathbf{F}$, show that the momentum of the particle is given by $\mathbf{p} = \mathbf{p}_0 + (mk/L)\hat{\boldsymbol{\theta}}$, where $\mathbf{p}_0$ is a constant vector and L is the magnitude of the angular momentum.
(b) Hence show that the orbit in momentum space (the so-called hodograph) is a circle. Where is the center and what is the radius of the circle?
(c) Show that the magnitude of $\mathbf{p}_0$ is $(mk/L)e$, where e is the eccentricity. Sketch the orbit in momentum space for the various cases, $e = 0$, $0 < e < 1$, $e = 1$, $e > 1$, indicating for the last two cases which part of the circle is relevant.
(See: Arnold Sommerfeld, *Mechanics*, (Academic Press, New York, NY, 1952), trans. Martin O. Stern, p. 33, 40, 242; Harold Abelson, Andrea diSessa, and Lee Rudolph, "Velocity space and the geometry of planetary orbits," Am. J. Phys. **43**, 579-589 (1975).)

Solution

According to Newton's second law the momentum $\mathbf{p}$ of a particle moving in a central force field $-(k/r^2)\hat{\mathbf{r}}$ changes at a rate

$$\frac{d\mathbf{p}}{dt} = -\frac{k}{r^2}\hat{\mathbf{r}} = -\frac{mk}{mr^2\dot{\theta}}\dot{\theta}\hat{\mathbf{r}}.$$

In the second equality we have multiplied and divided by $m\dot{\theta}$. The reason for this is that we recognize

$$L = mr^2\dot{\theta}$$

as the magnitude of the angular momentum, and we also note that

$$\frac{d\hat{\boldsymbol{\theta}}}{dt} = -\dot{\theta}\hat{\mathbf{r}}.$$

The preceding equation becomes

$$\frac{d\mathbf{p}}{dt} = \frac{mk}{L}\frac{d\hat{\boldsymbol{\theta}}}{dt}.$$

Since the angular momentum L of a particle moving in a central force field is constant, we can integrate immediately to obtain

$$\mathbf{p} = \frac{mk}{L}\hat{\boldsymbol{\theta}} + \mathbf{p}_0$$

where $\mathbf{p}_0$ is a constant vector. To identify this vector, we take the θ-component of the equation. Since the θ-component of $\mathbf{p}$ is L/r, and the θ-component of $\mathbf{p}_0$ is $p_0 \cos\theta$ where θ is the angle between $\mathbf{p}_0$ and $\hat{\boldsymbol{\theta}}$, we find

$$\frac{L}{r} = \frac{mk}{L} + p_0 \cos\theta.$$

This can be written

$$\frac{L^2}{mkr} = 1 + \frac{p_0 L}{mk}\cos\theta,$$

which we recognize as the equation of a conic section with eccentricity $e = p_0 L/mk$. The magnitude of $\mathbf{p}_0$ is thus

$$p_0 = \frac{mk}{L}e.$$

We also recognize θ as the angle from pericenter, so the direction of the vector $\mathbf{p}_0$ is perpendicular to a line from the force center to pericenter.

Another way to obtain the magnitude of $\mathbf{p}_0$ is to write

$$p_0^2 = \left|\mathbf{p} - \frac{mk}{L}\hat{\boldsymbol{\theta}}\right|^2 = p^2 - \frac{2mk}{r} + \frac{m^2k^2}{L^2}.$$

In the second equality we have used the fact that the θ-component of **p** is L/r. The first two terms on the right are $2mE$ where E is the total energy. The equation thus becomes

$$p_0^2 = \frac{m^2k^2}{L^2}e^2$$

where $e = \sqrt{1 + \frac{2EL^2}{mk^2}}$ is the eccentricity expressed in terms of the energy and the angular momentum. This then returns us to our previous expression for p_0.

We now note that

$$|\mathbf{p} - \mathbf{p}_0| = \frac{mk}{L}.$$

This is the equation of a circle, in momentum space, with center at $\mathbf{p}_0$ and with radius mk/L. Since $p_0 = (mk/L)e$, the circle has its center at the origin for circular ($e = 0$) orbits (Fig. 1(a)), encloses the origin for elliptic ($0 < e < 1$) orbits (Fig. 1(b)), passes through the origin for parabolic ($e = 1$) orbits (Fig. 1(c)), and excludes the origin for hyperbolic ($e > 1$) orbits (Fig. 1(d)).

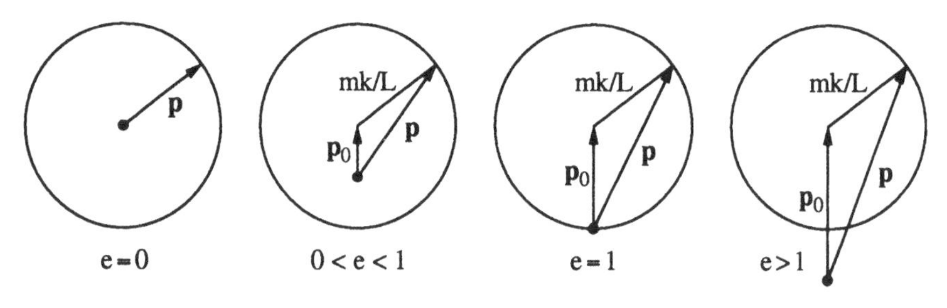

Ex. 1.11, Fig. 1(a) Ex. 1.11, Fig. 1(b) Ex. 1.11, Fig. 1(c) Ex. 1.11, Fig. 1(d)

For hyperbolic orbits we must also have, for an attractive force,

$$p^2 \geq 2mE = \left(\frac{mk}{L}\right)^2 \left(e^2 - 1\right).$$

This has the geometric significance shown in Fig. 2: points A and C, for which the equality holds, are the ends of tangent lines from the origin O to the circle; they correspond to those points on the orbit in real space at which $r \to \infty$. At point B the magnitude of the momentum is maximum; it corresponds to pericenter. Only the sector ABC of the circle in momentum space is relevant, for an attractive force.

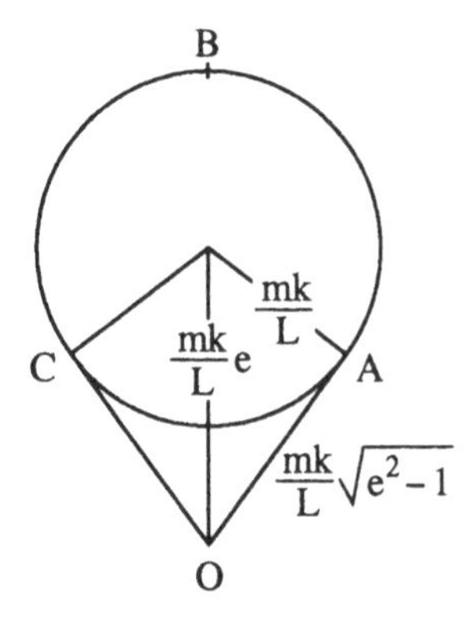

Ex. 1.11, Fig. 2

Exercise 1.12

Consider the motion of a particle in a central force field with potential $V = -k/r$. Since the force is central, the angular momentum $\mathbf{L} = \mathbf{r} \times \mathbf{p}$ is constant and the orbit lies in a plane passing through the force center and perpendicular to $\mathbf{L}$.
(a) Show that for the particular potential $V = -k/r$ there exists an additional vector quantity which is constant, the Laplace-Runge-Lenz vector

$$\mathbf{K} = \mathbf{p} \times \mathbf{L} - mk\hat{\mathbf{r}}.$$

Further show that $\mathbf{K} \cdot \mathbf{L} = 0$, so that $\mathbf{K}$ and $\mathbf{L}$ are perpendicular and thus $\mathbf{K}$ lies in the orbital plane. (Hint: if you've done Exercise 1.11, you need only show that $\mathbf{K} = \mathbf{p}_0 \times \mathbf{L}$).
(b) By taking the dot product of $\mathbf{K}$ with $\hat{\mathbf{r}}$ obtain the equation of the orbit

$$\frac{a(1-e^2)}{r} = 1 + e\cos\theta.$$

Hence find a and e in terms of K and L, and also find the direction that $\mathbf{K}$ points in the orbital plane.
(c) Express the energy $E = \dfrac{p^2}{2m} - \dfrac{k}{r}$ in terms of K and L.

Solution

(a) The time rate of change of the Laplace-Runge-Lenz vector **K** is

$$\begin{aligned}\frac{d\mathbf{K}}{dt} &= \frac{d\mathbf{p}}{dt}\times\mathbf{L}+\mathbf{p}\times\frac{d\mathbf{L}}{dt}-mk\frac{d}{dt}\left(\frac{\mathbf{r}}{r}\right)\\ &= -\frac{k}{r^3}\mathbf{r}\times\mathbf{L}+0-mk\left(\frac{\mathbf{v}}{r}-\frac{\mathbf{r}\mathbf{r}\cdot\mathbf{v}}{r^3}\right)\\ &= -\frac{k}{r^3}\mathbf{r}\times(\mathbf{r}\times\mathbf{p})-\frac{k}{r^3}(r^2\mathbf{p}-\mathbf{r}\mathbf{r}\cdot\mathbf{p})\\ &= 0\,,\end{aligned}$$

with the final equality following from the vector identity $\mathbf{a}\times(\mathbf{b}\times\mathbf{c})=(\mathbf{a}\cdot\mathbf{c})\mathbf{b}-(\mathbf{a}\cdot\mathbf{b})\mathbf{c}$. So the Laplace-Runge-Lenz vector **K** is constant in time. Further, we have

$$\mathbf{K}\cdot\mathbf{L}=(\mathbf{p}\times\mathbf{L})\cdot\mathbf{L}-(mk/r)\mathbf{r}\cdot(\mathbf{r}\times\mathbf{p})=0,$$

so the Laplace-Runge-Lenz vector **K** and the angular momentum vector **L** are perpendicular to one another.

(b) Now look at $\mathbf{K}\cdot\mathbf{r}=(\mathbf{p}\times\mathbf{L})\cdot\mathbf{r}-mkr$. This becomes $Kr\cos\theta=L^2-mkr$ where θ is the angle between (the constant) **K** and **r**. This, in turn, gives

$$\frac{L^2}{mkr}=1+\frac{K}{mk}\cos\theta,$$

which is the equation of a conic section with $L^2/mk=a(1-e^2)$ and $K/mk=e$, and with θ the angle from pericenter. The Laplace-Runge-Lenz vector **K** thus points from the force center towards pericenter.

Another way to obtain these results is as follows. Exercise 1.11 shows that

$$\mathbf{p}=\frac{mk}{L}\hat{\boldsymbol{\theta}}+\mathbf{p}_0$$

where $\mathbf{p}_0$ is a constant vector which is perpendicular to a line from the force center to pericenter. With this and with $\mathbf{L}=L\hat{\mathbf{z}}$ (switching temporarily to cylindrical coordinates) the Laplace-Runge-Lenz vector becomes

$$\mathbf{K}=\left(\frac{mk}{L}\hat{\boldsymbol{\theta}}+\mathbf{p}_0\right)\times\left(L\hat{\mathbf{z}}\right)-mk\hat{\mathbf{r}}=\mathbf{p}_0\times\mathbf{L},$$

since $\hat{\boldsymbol{\theta}}\times\hat{\mathbf{z}}=\hat{\mathbf{r}}$. Thus, we again see that **K** is a constant vector which points from the force center towards pericenter.

(c) The square of the length of the Laplace-Runge-Lenz vector is

$$\begin{aligned}K^2 &= |\mathbf{p}\times\mathbf{L}|^2 - (2mk/r)\mathbf{r}\cdot\mathbf{p}\times\mathbf{L} + m^2k^2\\ &= p^2L^2 - (2mk/r)L^2 + m^2k^2\\ &= 2mL^2E + m^2k^2\end{aligned}$$

where $E = p^2/2m - k/r$ is the total energy. So the energy, expressed in terms of K and L, is

$$E = \frac{1}{2mL^2}\left(K^2 - m^2k^2\right).$$

Exercise 1.13

Consider the motion of a particle of mass m in a central force field with potential

$$V = -\frac{k}{r} + \frac{h}{r^2}.$$

(a) Show that the equation for the orbit can be put in the form

$$\frac{a(1-e^2)}{r} = 1 + e\cos\alpha\theta,$$

and find a, e, and α in terms of the energy E and angular momentum L of the particle, and the parameters k and h of the potential.
(b) Show that this represents a precessing ellipse, and derive an expression for the average rate of precession in terms of the dimensionless quantity $\eta = h/ka$.
(c) The perihelion of Mercury precesses at the rate of 40″ of arc per century, after all known planetary perturbations are taken into account. What value of η would lead to this result? The eccentricity of Mercury's orbit is 0.206 and its period is 0.24 years.

Solution

(a) The orbit equation is given by

$$\theta = \int_{r_0}^{r}\frac{dr}{r^2\sqrt{\dfrac{2mE}{L^2} - \dfrac{1}{r^2} - \dfrac{2mV(r)}{L^2}}} = \int_{r_0}^{r}\frac{dr}{r^2\sqrt{\dfrac{2mE}{L^2} - \dfrac{\alpha^2}{r^2} + \dfrac{2mk}{L^2r}}}.$$

where $\alpha^2 = 1 + 2mh/L^2$. To do the r-integration, we first set $u = 1/r$, $du = -dr/r^2$ to give

$$\theta = -\int_{u_0}^{u} \frac{du}{\sqrt{\frac{2mE}{L^2} + \frac{2mk}{L^2}u - \alpha^2 u^2}} = -\frac{1}{\alpha}\int_{u_0}^{u} \frac{du}{\sqrt{\frac{m^2k^2}{\alpha^4 L^4}\left(1 + \frac{2\alpha^2 L^2 E}{mk^2}\right) - \left(u - \frac{mk}{\alpha^2 L^2}\right)^2}}.$$

We then set

$$u - \frac{mk}{\alpha^2 L^2} = \frac{mk}{\alpha^2 L^2}\sqrt{1 + \frac{2\alpha^2 L^2 E}{mk^2}}\cos A, \quad du = -\frac{mk}{\alpha^2 L^2}\sqrt{1 + \frac{2\alpha^2 L^2 E}{mk^2}}\sin A\, dA.$$

Integration gives $\theta = A/\alpha$, which leads to the orbit equation

$$u = \frac{1}{r} = \frac{mk}{\alpha^2 L^2}\left[1 + \sqrt{1 + \frac{2\alpha^2 L^2 E}{mk^2}}\cos\alpha\theta\right].$$

This has the form

$$\frac{a(1-e^2)}{r} = 1 + e\cos\alpha\theta$$

with $a(1-e^2) = \frac{\alpha^2 L^2}{mk}$, $e = \sqrt{1 + \frac{2\alpha^2 L^2 E}{mk^2}}$, and thus $a = -\frac{k}{2E}$.

(b) Pericenter occurs at $\cos\alpha\theta = 1$; that is, at $\theta = 0, 2\pi/\alpha, 4\pi/\alpha, \cdots$. The precession of pericenter per period is thus $2\pi(1/\alpha - 1)$, and the average rate of precession is

$$\text{Rate} = \frac{2\pi}{\tau}\left(\frac{1}{\alpha} - 1\right).$$

We have seen that

$$\alpha^2 = 1 + \frac{2mh}{L^2} = 1 + \frac{2h\alpha^2}{ka(1-e^2)},$$

so

$$\frac{1}{\alpha} = \sqrt{1 - \frac{2\eta}{1-e^2}}$$

where $\eta = h/ka$ is a dimensionless measure of the strength of the $1/r^2$ term in the potential. The rate of precession becomes

$$\text{Rate} = \frac{2\pi}{\tau}\left(\sqrt{1-\frac{2\eta}{1-e^2}}-1\right) \approx -\frac{2\pi}{\tau}\left(\frac{\eta}{1-e^2}\right).$$

(c) Applying this to Mercury, we have

$$\frac{1}{100}\frac{40}{60\times 60}\frac{\pi}{180} \approx \frac{2\pi}{0.24}\frac{\eta}{1-(0.206)^2}$$

which requires $\eta \approx 7\times 10^{-8}$.

Exercise 1.14

A particle of mass m moves in a 3D isotropic harmonic oscillator potential well

$$V = \tfrac{1}{2}m\omega^2 r^2$$

where ω, the angular frequency, is a constant.
(a) Show that the equation for the orbit has the form

$$\frac{L^2}{mE}\frac{1}{r^2} = 1 + \sqrt{1-\frac{\omega^2 L^2}{E^2}}\cos 2(\theta-\theta_0)$$

where E is the energy and L is the angular momentum.
(b) Show that this represents an ellipse with geometric center at the force center, and express the energy and angular momentum in terms of the semi-major axis a and eccentricity e of the ellipse. (Ans. $E = \tfrac{1}{2}m\omega^2(a^2+b^2)$ and $L = m\omega ab$ where $b = a\sqrt{1-e^2}$ is the semi-minor axis)
(c) Show that the period is $\tau = 2\pi/\omega$ independent of the energy and angular momentum, and that the radius is given as a function of time by

$$r^2 = \frac{E}{m\omega^2}\left[1-\sqrt{1-\frac{\omega^2 L^2}{E^2}}\cos 2\omega(t-t_0)\right].$$

Solution

(a) The orbit equation is given by

$$\theta = \int_{r_0}^{r} \frac{dr}{r^2\sqrt{\frac{2mE}{L^2} - \frac{1}{r^2} - \frac{2mV(r)}{L^2}}} = \int_{r_0}^{r} \frac{dr}{r^2\sqrt{\frac{2mE}{L^2} - \frac{1}{r^2} - \frac{m^2\omega^2 r^2}{L^2}}}.$$

To do the r-integration, we first set $u = 1/r$, $du = -dr/r^2$ to obtain

$$\theta = -\int_{u_0}^{u} \frac{u\,du}{\sqrt{\frac{2mE}{L^2}u^2 - u^4 - \frac{m^2\omega^2}{L^2}}} = -\int_{u_0}^{u} \frac{u\,du}{\sqrt{\frac{m^2E^2}{L^2}\left(1 - \frac{\omega^2L^2}{E^2}\right) - \left(u^2 - \frac{mE}{L^2}\right)^2}}.$$

We then set

$$u^2 - \frac{mE}{L^2} = \frac{mE}{L^2}\sqrt{1 - \frac{\omega^2L^2}{E^2}}\cos A, \qquad 2u\,du = -\frac{mE}{L^2}\sqrt{1 - \frac{\omega^2L^2}{E^2}}\sin A\,dA.$$

Integration gives $\theta = A/2$, which leads to the orbit equation

$$u^2 = \frac{1}{r^2} = \frac{mE}{L^2}\left[1 + \sqrt{1 - \frac{\omega^2L^2}{E^2}}\cos 2\theta\right].$$

(b) To identify this, we set $\cos 2\theta = \cos^2\theta - \sin^2\theta$ in the orbit equation and multiply it through by L^2r^2/mE. Recalling that the cartesian coordinates are given by $x = r\cos\theta$, $y = r\sin\theta$, we can rewrite the orbit equation in the form

$$\frac{L^2}{mE} = \left[1 + \sqrt{1 - \frac{\omega^2L^2}{E^2}}\right]x^2 + \left[1 - \sqrt{1 - \frac{\omega^2L^2}{E^2}}\right]y^2.$$

We recognize this as the equation of an ellipse with geometric center at the origin, and with semi-major axis a (in the y-direction) and semi-minor axis b (in the x-direction) given by

$$a^2 = \frac{L^2/mE}{1 - \sqrt{1 - \omega^2L^2/E^2}}, \qquad b^2 = \frac{L^2/mE}{1 + \sqrt{1 - \omega^2L^2/E^2}}.$$

These give $a^2b^2 = L^2/m^2\omega^2$, so $L = m\omega ab$. Further, $1/a^2 + 1/b^2 = 2mE/L^2$, so $E = \frac{1}{2}m\omega^2(a^2 + b^2)$.

(c) The rate at which the radius vector sweeps out area is

$$\frac{dA}{dt} = \frac{L}{2m}.$$

Since the total area enclosed by the ellipse is $A = \pi ab$ and the angular momentum is $L = m\omega ab$, this becomes

$$\frac{\pi ab}{\tau} = \frac{\omega ab}{2},$$

which gives the period $\tau = 2\pi/\omega$. Note that it is independent of the energy and angular momentum.

The time dependence of r is obtained from

$$\sqrt{\frac{2}{m}}\,t = \int_{r_0}^{r} \frac{dr}{\sqrt{E - L^2/2mr^2 - \frac{1}{2}m\omega^2 r^2}} = \int_{r_0}^{r} \frac{r\,dr}{\sqrt{Er^2 - L^2/2m - \frac{1}{2}m\omega^2 r^4}}.$$

To perform the r-integration, we first set $s = r^2$, $ds = 2r\,dr$ to give

$$2\omega t = \int_{r_0}^{r} \frac{ds}{\sqrt{\frac{E^2}{m^2\omega^4}\left(1 - \frac{\omega^2 L^2}{E^2}\right) - \left(s - \frac{E}{m\omega^2}\right)^2}}.$$

We then set

$$s - \frac{E}{m\omega^2} = -\frac{E}{m\omega^2}\sqrt{1 - \frac{\omega^2 L^2}{E^2}}\cos T \quad \text{and} \quad ds = \frac{E}{m\omega^2}\sqrt{1 - \frac{\omega^2 L^2}{E^2}}\sin T\,dT.$$

Integration gives $2\omega t = T$, so the time dependence of r is given by

$$s = r^2 = \frac{E}{m\omega^2}\left[1 - \sqrt{1 - \frac{\omega^2 L^2}{E^2}}\cos 2\omega t\right].$$

Exercise 1.15

A small meteor approaches the earth with impact parameter b and velocity v_∞ at infinity. Show that the meteor will strike the earth if

$$b < a\sqrt{1 + (v_0/v_\infty)^2}$$

where a is the radius and v_0 is the "escape velocity" for the earth.

Solution

One way to do this exercise is to use conservation of angular momentum and energy,

$$L = mv_a a = mv_\infty b,$$
$$E = \tfrac{1}{2}mv_a^2 - k/a = \tfrac{1}{2}mv_\infty^2 .$$

In the middle expressions we have assumed that the meteor is just grazing the earth at radius a and with speed v_a. Eliminating v_a, we find

$$\frac{1}{2}mv_\infty^2 \frac{b^2}{a^2} - \frac{k}{a} = \frac{1}{2}mv_\infty^2.$$

The escape velocity v_0 for the earth is given by

$$\frac{1}{2}mv_0^2 = \frac{k}{a}.$$

Using this to eliminate k/a, we obtain

$$\frac{b^2}{a^2} = 1 + \frac{v_0^2}{v_\infty^2}$$

as required.

Another way to do the exercise is to use the relation

$$\tfrac{1}{2}m\dot{r}^2 + V_{eff}(r) = E.$$

For an orbit which just grazes the earth, $r = a$, $\dot{r} = 0$, and thus

$$V_{eff}(a) = \frac{L^2}{2ma^2} - \frac{k}{a} = E.$$

Setting $L = mv_\infty b$ and $E = \tfrac{1}{2}mv_\infty^2$ then returns us to the preceding approach.

Exercise 1.16

(a) Find the relation between the scattering angle Θ and the impact parameter b for scattering from a hard sphere of radius a (for which "angle of incidence = angle of reflection").
(b) Use your result to obtain the differential scattering cross section $d\sigma/d\Omega$. Integrate to find the total scattering cross section $\sigma = \int (d\sigma/d\Omega)d\Omega$, where the integration extends over the whole solid angle.

Solution

(a)

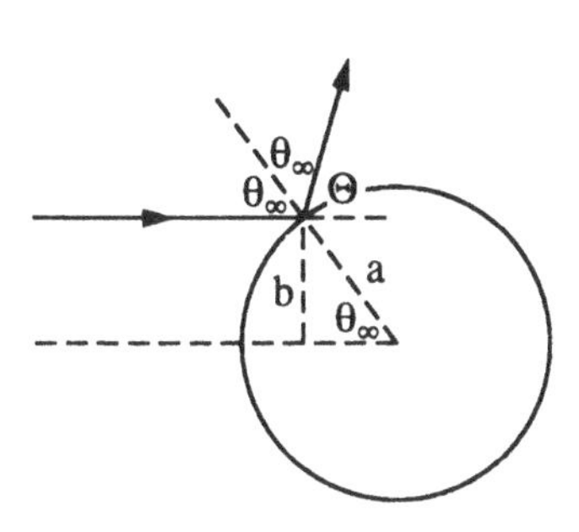

Ex. 1.16, Fig. 1

From Fig. 1 the impact parameter is $b = a\sin\theta_\infty$ and the scattering angle is $\Theta = \pi - 2\theta_\infty$. Eliminating θ_∞, we find the relation between the impact parameter and the scattering angle

$$b = a\cos(\Theta/2).$$

(b) The differential scattering cross section for hard sphere scattering is then

$$\frac{d\sigma}{d\Omega} = \frac{b}{\sin\Theta}\left|\frac{db}{d\Theta}\right| = \frac{a\cos(\Theta/2)}{\sin\Theta}\frac{a}{2}\sin\frac{\Theta}{2} = \frac{a^2}{4}.$$

Note that it is independent of the scattering angle. The total scattering cross section is

$$\sigma = \int \frac{d\sigma}{d\Omega}d\Omega = \frac{a^2}{4}4\pi = \pi a^2$$

and, as might be expected, equals the cross-sectional area of the sphere.

Exercise 1.17

(a) Show that a particle of energy E is refracted in going from a region in which the potential is zero to a region in which the potential is $-V_1$, the angle of incidence θ_0 and the angle of refraction θ_1 being related by Snell's law

$$\frac{\sin\theta_0}{\sin\theta_1} = n$$

where angles are measured from the normal and $n = \sqrt{1 + V_1/E}$ is the index of refraction.
(b) Use Snell's law to show that a particle incident at impact parameter b on an attractive square well potential

$$\begin{aligned} V(x) &= -V_1 \quad \text{for } r < a \\ V(x) &= 0 \quad \text{for } r > a \end{aligned}$$

is scattered through an angle Θ given by

$$\frac{b^2}{a^2} = \frac{n^2 \sin^2 \Theta/2}{n^2 + 1 - 2n\cos\Theta/2}.$$

In particular, show that for small impact parameters ($b << a$) the scattered particles are brought to a focus a distance $f \approx \left(\frac{n}{n-1}\right)\left(\frac{a}{2}\right)$ from the force center.
(c) Find the differential scattering cross section $d\sigma/d\Omega$.

Solution

(a)

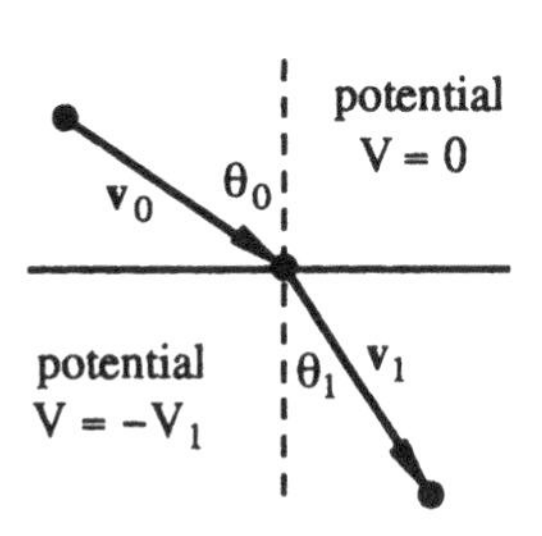

Ex. 1.17, Fig. 1

A particle, moving at speed v_0 in a region in which the potential is zero, is incident at angle θ_0 to the normal on the surface of a region in which the potential is $-V_1$. As the particle enters the region it encounters an inward impulsive force. The velocity of the particle parallel to the surface is unaffected, but the velocity perpendicular to the surface is increased and hence the direction of motion changes; the particle is "refracted." Let v_1 be the speed of the particle and θ_1 the angle the direction of motion makes with the normal in the region in which the potential is $-V_1$ (Fig. 1). From the fact that the tangential component of the velocity is constant we have

$$v_1 \sin\theta_1 = v_0 \sin\theta_0,$$

and from conservation of energy we have

$$\tfrac{1}{2}mv_1^2 - V_1 = \tfrac{1}{2}mv_0^2 = E.$$

Combining these relations, we obtain

$$\frac{\sin\theta_0}{\sin\theta_1} = \frac{v_1}{v_0} = \sqrt{1 + \frac{V_1}{E}}.$$

This has the form of Snell's law with $n = \sqrt{1 + V_1/E}$ being the "index of refraction." Note, however, that particles speed up on entering a region with "index of refraction" $n > 1$ whereas waves slow down.

(b) Suppose now that the particle is incident at impact parameter b on an attractive square well potential (Fig. 2).

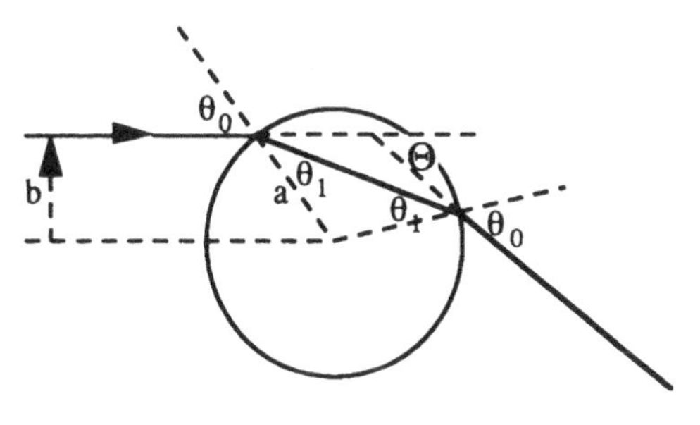

Ex. 1.17, Fig. 2

The incident angle θ_0 is given by

$$\sin\theta_0 = b/a$$

and the refracted angle θ_1 by

$$\sin\theta_1 = b/na.$$

The scattering angle is

$$\Theta = 2(\theta_0 - \theta_1),$$

and thus

$$\cos(\Theta/2) = \cos\theta_0 \cos\theta_1 + \sin\theta_0 \sin\theta_1$$
$$= \sqrt{1 - \frac{b^2}{a^2}}\sqrt{1 - \frac{b^2}{n^2a^2}} + \frac{b^2}{na^2}.$$

The scattering angle ranges from zero at impact parameter zero to a maximum of $2\cos^{-1}(1/n)$ at impact parameter a. We can instead express the impact parameter in terms of the scattering angle. Rearranging and squaring the preceding equation, we find

$$\left(\cos\frac{\Theta}{2} - \frac{b^2}{na^2}\right)^2 = \left(1 - \frac{b^2}{a^2}\right)\left(1 - \frac{b^2}{n^2a^2}\right)$$

which simplifies to

$$\frac{b^2}{a^2} = \frac{n^2 \sin^2 \Theta/2}{n^2 + 1 - 2n\cos\Theta/2}$$

(with $0 < \Theta < 2\cos^{-1}(1/n)$). For small impact parameter this becomes

$$\Theta \approx 2\left(\frac{b}{a}\right)\left(1 - \frac{1}{n}\right).$$

Let f, the "focal length," be the distance from the force center to the point at which the scattered particle crosses the axis of symmetry. To find this distance, we drop perpendiculars from the force center onto the lines of the incoming and outgoing trajectories. The length of that onto the incoming trajectory is the impact parameter b, and symmetry implies that the length of these perpendiculars are equal (Fig. 3).

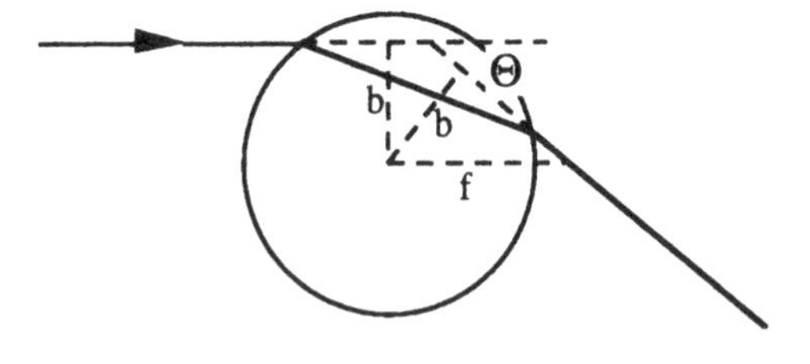

Ex. 1.17, Fig. 3

We see that

$$f = b/\sin\Theta,$$

and thus

$$f = \frac{na}{2\cos\Theta/2\sqrt{n^2+1-2n\cos\Theta/2}}.$$

For small impact parameter the focal length becomes

$$f \approx \frac{na}{2(n-1)}$$

and is independent of the impact parameter. The particles are then brought to a focus a distance f from the force center. These remarks on the focal length hold only for weak attractive potentials. Otherwise, the geometry changes with the particle crossing the axis inside the potential region.

(c) The differential scattering cross section is given by

$$\frac{d\sigma}{d\Omega} = \frac{b}{\sin\Theta}\left|\frac{db}{d\Theta}\right|.$$

Since

$$\begin{aligned}\frac{2b}{a^2}\frac{db}{d\Theta} &= \frac{n^2\sin(\Theta/2)\cos(\Theta/2)}{n^2+1-2n\cos(\Theta/2)} - \frac{n^3\sin^3(\Theta/2)}{(n^2+1-2n\cos(\Theta/2))^2} \\ &= \frac{n^2\sin(\Theta/2)(n\cos(\Theta/2)-1)(n-\cos(\Theta/2))}{(n^2+1-2n\cos(\Theta/2))^2},\end{aligned}$$

we have

$$\frac{d\sigma}{d\Omega} = \frac{n^2a^2}{4\cos(\Theta/2)}\frac{(n\cos(\Theta/2)-1)(n-\cos(\Theta/2))}{(n^2+1-2n\cos(\Theta/2))^2}.$$

For large index of refraction (strong *attractive* potential) the differential scattering cross section becomes

$$\frac{d\sigma}{d\Omega} \approx \frac{a^2}{4}$$

and is the same as that for scattering from a hard sphere (strong *repulsive* potential); see Exercise 1.16. To understand this, note that the outgoing trajectories in the two cases are in exactly opposite directions for a given impact parameter (Fig. 4), so the cross-sections, which are invariant under inversion, are equal.

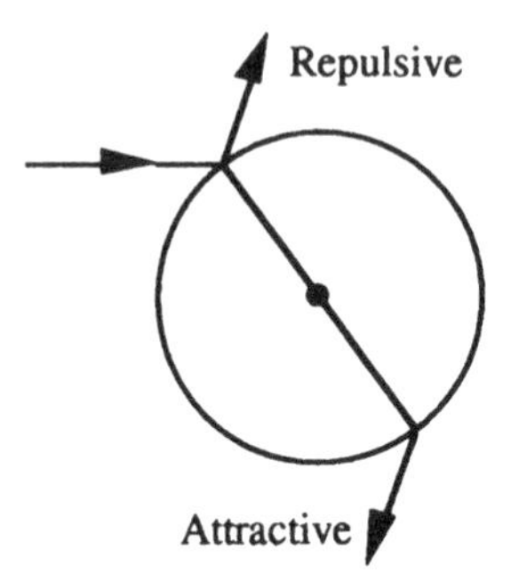

Ex. 1.17, Fig. 4

Exercise 1.18

(a) Show that

$$\frac{r_0}{r} = \cos\alpha\theta$$

is the equation of the orbit for a particle moving in a repulsive potential $V(r) = k/r^2$, determining α and r_0 in terms of the energy and angular momentum.
(Ans. $\alpha = \sqrt{1 + \frac{2mk}{L^2}}$, $r_0 = \frac{\alpha L}{\sqrt{2mE}}$)
(b) Show that the impact parameter b and scattering angle Θ are related by

$$b^2 = \frac{k}{E}\frac{(\pi - \Theta)^2}{\Theta(2\pi - \Theta)}.$$

(c) Show that the differential scattering cross section is given by

$$\frac{d\sigma}{d\Omega} = \frac{\pi^2 k}{E\sin\Theta}\frac{\pi - \Theta}{\Theta^2(2\pi - \Theta)^2}.$$

Solution

(a) The orbit equation is given by

$$\theta = \int_{r_0}^{r} \frac{dr}{r^2\sqrt{\dfrac{2mE}{L^2} - \dfrac{1}{r^2} - \dfrac{2mV(r)}{L^2}}} = \int_{r_0}^{r} \frac{dr}{r^2\sqrt{\dfrac{2mE}{L^2} - \dfrac{\alpha^2}{r^2}}}$$

where $\alpha^2 = 1 + 2mk/L^2$. To do the r-integration, we first set $u = 1/r$, $du = -dr/r^2$ to obtain

$$\theta = -\frac{1}{\alpha}\int_{u_0}^{u} \frac{du}{\sqrt{2mE/\alpha^2L^2 - u^2}}.$$

We then set

$$u = (\sqrt{2mE}/\alpha L)\cos A, \quad du = -(\sqrt{2mE}/\alpha L)\sin A\, dA.$$

Integration gives $\theta = A/\alpha$, so the equation of the orbit is

$$u = 1/r = (\sqrt{2mE}/\alpha L)\cos\alpha\theta;$$

that is, $r_0/r = \cos\alpha\theta$ where $r_0 = \alpha L/\sqrt{2mE}$.

(b) Pericenter is at $\theta = 0$, and for $r \to \infty$ the angle $\alpha\theta \to \pm\pi/2$, so

$$\theta_\infty = \pi/2\alpha.$$

Expressing θ_∞ in terms of the scattering angle Θ by using $\Theta = \pi - 2\theta_\infty$, and α in terms of the impact parameter b by using $L = b\sqrt{2mE}$, we find

$$\frac{1}{2}(\pi - \Theta) = \frac{\pi}{2\sqrt{1 + k/Eb^2}}.$$

Solving for b, we obtain

$$b^2 = \frac{k}{E}\frac{(\pi - \Theta)^2}{\Theta(2\pi - \Theta)}.$$

(c) The differential scattering cross section is given by

$$\frac{d\sigma}{d\Omega} = \frac{b}{\sin\Theta}\left|\frac{db}{d\Theta}\right|.$$

Since

$$2b\frac{db}{d\Theta} = -\frac{2\pi^2 k}{E}\frac{\pi - \Theta}{\Theta^2(2\pi - \Theta)^2},$$

we have

$$\frac{d\sigma}{d\Omega} = \frac{\pi^2 k}{E\sin\Theta}\frac{\pi - \Theta}{\Theta^2(2\pi - \Theta)^2}.$$

For small scattering angles this becomes

$$\frac{d\sigma}{d\Omega} \approx \frac{\pi k}{4E}\frac{1}{\Theta^3},$$

whereas for $\Theta \to \pi$ it becomes

$$\frac{d\sigma}{d\Omega} \to \frac{k}{\pi^2 E}.$$

CHAPTER II

THE PRINCIPLE OF VIRTUAL WORK AND D'ALEMBERT'S PRINCIPLE

Exercise 2.01

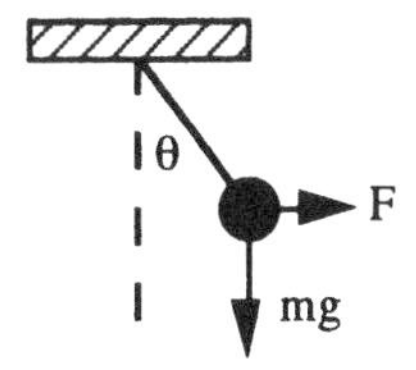

Use d'Alembert's principle to find the condition of static equilibrium.

Solution

Imagine a virtual displacement in which the angle θ increases by a small amount $\delta\theta$. The mass moves horizontally a distance $\delta x = \ell\cos\theta\delta\theta$, and the applied force F does work $F\delta x = F\ell\cos\theta\delta\theta$. The mass moves vertically upwards a distance $\delta y = \ell\sin\theta\delta\theta$, and the applied force mg does work $-mg\delta y = -mg\ell\sin\theta\delta\theta$. There are no inertial forces, so d'Alembert's principle gives

$$F\ell\cos\theta\,\delta\theta - mg\ell\sin\theta\,\delta\theta = 0,$$

which simplifies to

$$F = mg\tan\theta.$$

This is the condition of static equilibrium.

Exercise 2.02

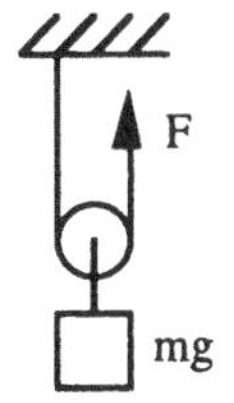

Use d'Alembert's principle to find the condition of static equilibrium.

Solution

Imagine a virtual displacement in which the end of the string is raised a distance δx and the weight is thus raised a distance $\delta x/2$. The applied force F does work $F\delta x$ and the applied force mg does work $-mg\delta x/2$. There are no inertial forces, so d'Alembert's principle gives

$$F\delta x - mg\delta x/2 = 0.$$

This yields $F = mg/2$, the condition of static equilibrium.

Exercise 2.03

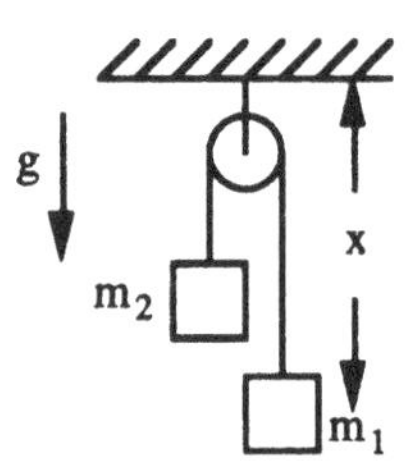

Use d'Alembert's principle to find the acceleration of m_1.

Solution

Imagine a virtual displacement in which m_1 moves downwards a distance δx and m_2 moves upwards a distance δx. The applied force, gravity, does work $m_1 g\delta x$ on m_1 and $-m_2 g\delta x$ on m_2. The acceleration of m_1 is $\ddot{x}$ downwards, and that of m_2 is $\ddot{x}$ upwards. The inertial force on m_1 is $m_1\ddot{x}$ upwards and the inertial work done on m_1 is $-m_1\ddot{x}\delta x$. The inertial force on m_2 is $m_2\ddot{x}$ downwards and the inertial work done on m_2 is $-m_2\ddot{x}\delta x$. D'Alembert's principle gives

$$(m_1 - m_2)g\delta x - (m_1 + m_2)\ddot{x}\delta x = 0,$$

so the acceleration of m_1 is

$$\ddot{x} = \frac{m_1 - m_2}{m_1 + m_2} g.$$

Exercise 2.04

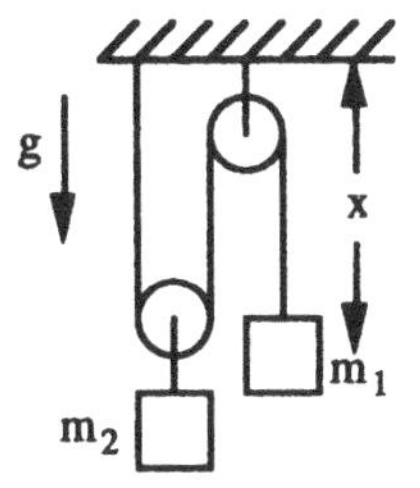

Use d'Alembert's principle to find the acceleration of m_1.

Solution

Imagine a virtual displacement in which m_1 moves downwards a distance δx and m_2 thus moves upwards a distance $\delta x/2$. The applied force, gravity, does work $m_1 g \delta x$ on m_1 and $-m_2 g \delta x/2$ on m_2. The acceleration of m_1 is $\ddot{x}$ downwards, and that of m_2 is $\ddot{x}/2$ upwards. The inertial force on m_1 is $m_1\ddot{x}$ upwards, and that on m_2 is $m_2\ddot{x}/2$ downwards. These inertial forces do work $-m_1\ddot{x}\delta x$ on m_1 and $-(m_2\ddot{x}/2)(\delta x/2)$ on m_2. D'Alembert's principle gives

$$(m_1 - \tfrac{1}{2}m_2)g\delta x - (m_1 + \tfrac{1}{4}m_2)\ddot{x}\delta x = 0,$$

so the acceleration of m_1 is

$$\ddot{x} = \frac{m_1 - \frac{1}{2}m_2}{m_1 + \frac{1}{4}m_2}g.$$

Exercise 2.05

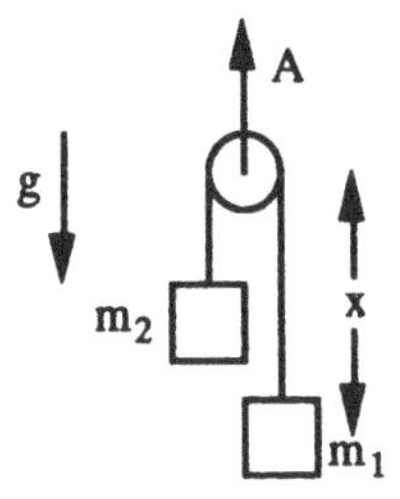

Use d'Alembert's principle to find the acceleration of m_1. Note that in this case the pulley has an upward acceleration A. "Acceleration" means "acceleration relative to the earth."

Solution

Imagine a virtual displacement in which m_1 moves downwards a distance δx and m_2 moves upwards a distance δx. The applied force, gravity, does work $m_1 g\delta x$ on m_1 and $-m_2 g\delta x$ on m_2. The acceleration of m_1 is $(\ddot{x} - A)$ downwards, and that of m_2 is $(\ddot{x} + A)$ upwards. The inertial force on m_1 is $m_1(\ddot{x} - A)$ upwards, and that on m_2 is $m_2(\ddot{x} + A)$ downwards. These inertial forces do work $-m_1(\ddot{x} - A)\delta x$ on m_1 and $-m_2(\ddot{x} + A)\delta x$ on m_2. D'Alembert's principle gives

$$m_1 g\delta x - m_2 g\delta x - m_1(\ddot{x} - A)\delta x - m_2(\ddot{x} + A)\delta x = 0,$$

which simplifies to

$$\ddot{x} = \frac{m_1 - m_2}{m_1 + m_2}(g + A).$$

In the frame of the pulley this system, and thus the downward acceleration $\ddot{x}$ of m_1, is the same as an "Atwood's machine" (Exercise 2.03) in a gravitational field $g + A$. In the frame of the earth the downward acceleration of m_1 is

$$\ddot{x} - A = \frac{(m_1 - m_2)g - 2m_2 A}{m_1 + m_2}.$$

Suppose $m_1 > m_2$. Then, if the upward acceleration A of the pulley is small, m_1 accelerates downwards and m_2 accelerates upwards. However, if

$$A > \frac{m_1 - m_2}{2m_2} g,$$

both masses accelerate upwards.

Exercise 2.06

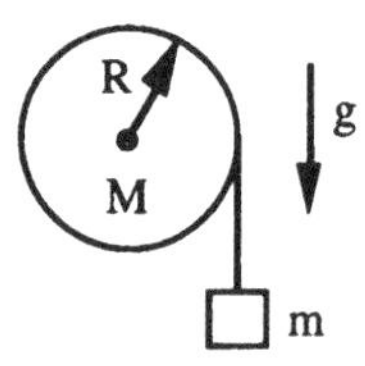

A mass m is attached to a light cord which wraps around a frictionless pulley of mass M, radius R, and moment of inertia $I = \int r^2 dM$. Gravity g acts vertically downwards. Use d'Alembert's principle to find the acceleration of m.

Solution

Imagine a virtual displacement in which m moves downwards a distance δs. The applied force, gravity, does work $mg\delta s$ on m. The inertial force on m is $m\ddot{s}$ upwards and the work it does is $-m\ddot{s}\delta s$. There is no applied work done on the (uniform) pulley, but there is inertial work. To find this, consider a little piece dM of the pulley which is at a distance r from the axle. This piece undergoes a virtual displacement, in the angular direction, of $(r/R)\delta s$. The acceleration of dM, in the angular direction, is $(r/R)\ddot{s}$ so the inertial force on dM, in the angular direction, is $-dM(r/R)\ddot{s}$. The inertial work done on dM is thus $-dM(r/R)^2\ddot{s}\delta s$ (dM also has a radial centripetal acceleration, but this does not contribute to the inertial work). Summing, we have for the total inertial work done on the pulley

$$-\int r^2\, dM/R^2\, \ddot{s}\delta s = -(I/R^2)\ddot{s}\delta s$$

where $I = \int r^2 dM$ is the moment of inertia of the pulley. This expression for the inertial work done in a fixed axis rotation can also be written in the useful general form $-I\alpha\delta\theta$ where $\alpha = \ddot{s}/R$ is the angular acceleration and $\delta\theta = \delta s/R$ is the angular displacement. Returning to the original problem, d'Alembert's principle now gives

$$mg\delta s - m\ddot{s}\delta s - (I/R^2)\ddot{s}\delta s = 0,$$

so the acceleration of m is

$$\ddot{s} = \frac{g}{1 + I/mR^2}.$$

Exercise 2.07

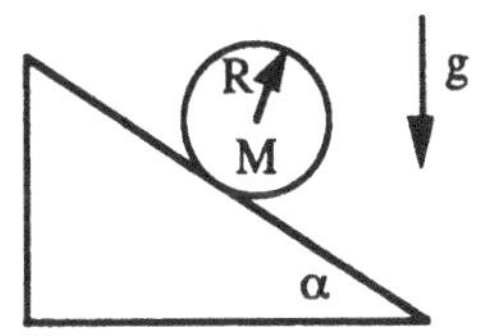

A cylinder of mass M, radius R, and moment of inertia $I = \int r^2 dM$ rolls without slipping down an inclined plane. Use d'Alembert's principle to find the acceleration of the cylinder.

Solution

Imagine a virtual displacement in which the cylinder rolls a small distance δs down the plane. The only applied force, gravity, does work

$$\delta W^{(applied)} = Mg\sin\alpha\,\delta s.$$

To find the inertial work is more difficult. In this virtual displacement the center of mass of the cylinder is translated δs and, because the cylinder rolls, it rotates about its center of mass through an angle $\delta\theta = \delta s/R$. A little piece dM of the cylinder at (r,θ) thus undergoes a displacement δs down the plane together with a displacement $r\delta\theta = r(\delta s/R)$ in the "angular direction." This latter displacement has components $-r(\delta s/R)\cos\theta$ down the plane and $r(\delta s/R)\sin\theta$ perpendicular to the plane (Fig. 1).

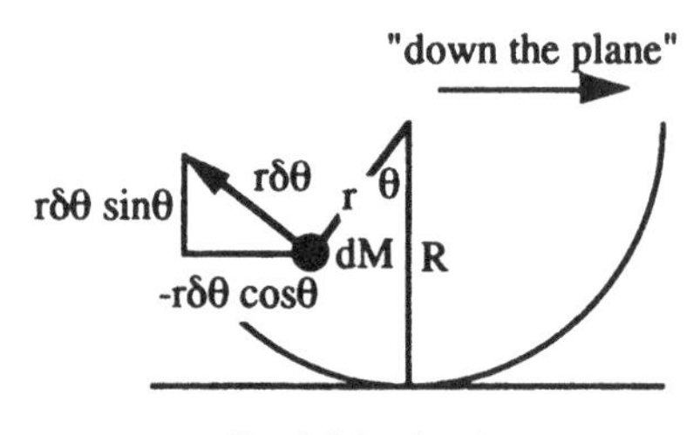

Ex. 2.07, Fig. 1

dM thus undergoes a net displacement $\delta s(1-(r/R)\cos\theta)$ down the plane and $\delta s(r/R)\sin\theta$ perpendicular to the plane. To find the inertial force on dM, we need its acceleration. The acceleration of the center of mass of the cylinder is $\ddot{s}$ down the plane, and the acceleration of dM with respect to the center of mass has components $r(\ddot{s}/R)$ in the "angular direction" and $-r(\dot{s}/R)^2$ (centripetal acceleration) in the "radial direction." Resolving these into their components parallel and perpendicular to the plane, we find the net acceleration of dM to be

$$\ddot{s} - r(\ddot{s}/R)\cos\theta + r(\dot{s}/R)^2\sin\theta \qquad \text{down the plane}$$

and

$$r(\ddot{s}/R)\sin\theta + r(\dot{s}/R)^2\cos\theta \qquad \text{perpendicular to the plane.}$$

A more analytical way to obtain these results is to note that the cartesian coordinates of dM are

$$x = s - r\sin\theta, \qquad y = R - r\cos\theta,$$

where x is "down the plane" and y is "perpendicular to the plane." Differentiating once, we obtain for dM a virtual displacement

$$\delta x = \delta s - r\cos\theta\,\delta\theta, \qquad \delta y = r\sin\theta\,\delta\theta.$$

Differentiating twice, we obtain for dM an acceleration

$$\ddot{x} = \ddot{s} - r\ddot{\theta}\cos\theta + r\dot{\theta}^2\sin\theta, \qquad \ddot{y} = r\ddot{\theta}\sin\theta + r\dot{\theta}^2\cos\theta.$$

The inertial force on dM is $-dM$ times the acceleration, so the inertial work done on dM is

$$\begin{aligned}&-dM[\ddot{s} - r(\ddot{s}/R)\cos\theta + r(\dot{s}/R)^2\sin\theta][\delta s(1-(r/R)\cos\theta]\\ &-dM[r(\ddot{s}/R)\sin\theta + r(\dot{s}/R)^2\cos\theta][\delta s(r/R)\sin\theta]\\ &= -dM\ddot{s}[1 - 2(r/R)\cos\theta + (r/R)^2]\delta s - dM(\dot{s}^2/R)(r/R)\sin\theta\,\delta s.\end{aligned}$$

On integration over the cylinder the angle-dependent terms go out and we obtain

$$\delta W^{(\text{inertial})} = -M\ddot{s}(1 + I/MR^2)\delta s$$

where $I = \int r^2 dM$ is the moment of inertia of the cylinder about its center of mass. This expression can be written in the general form $-M\ddot{s}\delta s - I\alpha\delta\theta$, where α is the angular acceleration of the cylinder and $\delta\theta$ is its angular displacement.

Applying d'Alembert's principle to the original problem, we find

$$Mg\sin\alpha\,\delta s - M\ddot{s}(1 + I/MR^2)\delta s = 0,$$

so the acceleration of the cylinder down the plane is

$$\ddot{s} = \frac{g\sin\alpha}{1 + I/MR^2}.$$

Exercise 2.08

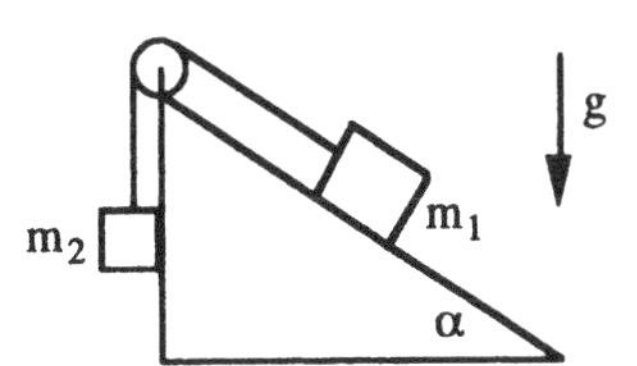

Use d'Alembert's principle to find the acceleration of m_1 down the (stationary) plane.

Solution

Imagine a virtual displacement in which m_1 moves down the plane a distance δs. The mass m_1 then undergoes a downward vertical displacement $\delta s\sin\alpha$ and the applied force, gravity, does work $m_1 g\delta s \sin\alpha$ on it. As well, m_2 moves vertically upwards a distance δs, and gravity does work $-m_2 g\delta s$ on it. If the acceleration of m_1 is " a" down the plane, the inertial force on it is $m_1 a$ up the plane, and the inertial work done on it is $-m_1 a\delta s$. The acceleration of m_2 is then " a" vertically upwards, the inertial force on it is $m_2 a$ vertically downwards, and the inertial work done on it is $-m_2 a\delta s$. D'Alembert's principle gives

$$m_1 g \sin\alpha\, \delta s - m_2 g \delta s - m_1 a \delta s - m_2 a \delta s = 0,$$

so the acceleration of m_1 down the plane is

$$a = \frac{m_1 \sin\alpha - m_2}{m_1 + m_2} g.$$

Exercise 2.09

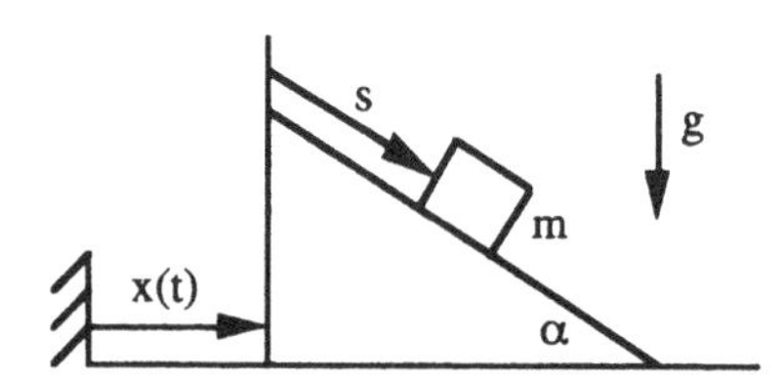

A block of mass m slides on a frictionless inclined plane, which is driven so that it moves horizontally, the displacement of the plane at time t being some known function $x(t)$. Use d'Alembert's principle to find the equation of motion of the block, taking as generalized coordinate the displacement s of the block down the plane. Note that the acceleration of the block is *not* "down the plane."

Solution

Imagine a virtual displacement in which s increases by δs. The mass m undergoes a downward vertical displacement $\delta s \sin\alpha$ and the applied force, gravity, does work $mg\delta s \sin\alpha$. The acceleration $\mathbf{a}$ of m is the vector sum of $\ddot{s}$ down the plane and $\ddot{x}$ horizontal (Fig. 1).

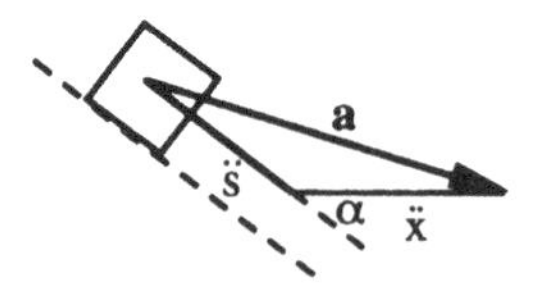

Ex. 2.09, Fig. 1

The inertial force on the mass is $-m\mathbf{a}$. We need the component of this in the direction of the virtual displacement, namely $-m(\ddot{s} + \ddot{x}\cos\alpha)$. The inertial work done is thus $-m(\ddot{s} - \ddot{x}\cos\alpha)\delta s$. D'Alembert's principle gives

$$mg\sin\alpha\,\delta s - m(\ddot{s} + \ddot{x}\cos\alpha)\delta s = 0,$$

so the equation of motion of the block is

$$\ddot{s} = g\sin\alpha - \ddot{x}\cos\alpha.$$

For $\ddot{x} = 0$ this reduces to the usual result for an inclined plane. For $\ddot{x} > g\tan\alpha$, however, the acceleration $\ddot{s}$ is negative and the mass m accelerates *up* the plane.

Exercise 2.10

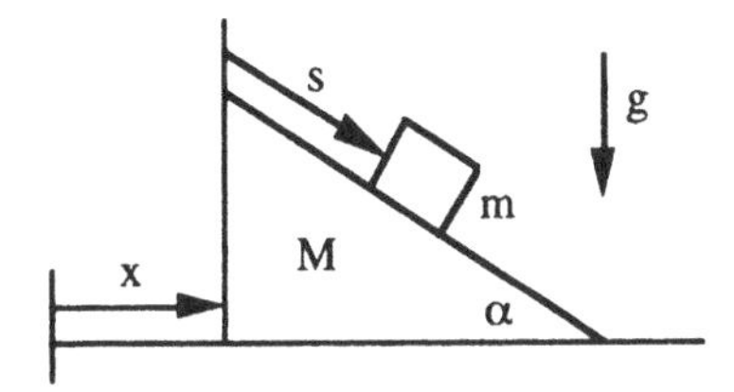

A block of mass m slides on a frictionless inclined plane of mass M which in turn is free to slide on a frictionless horizontal surface. Use d'Alembert's principle to find the equations of motion of the block and the plane, taking as generalized coordinates the displacement s of the block down the plane and the horizontal displacement x of the plane.

Solution

The system in this exercise has two degrees of freedom, and there are two independent virtual displacements. The first can be taken as in Exercise 2.09, an increase of s by δs. The application of d'Alembert's principle for this virtual displacement then proceeds as in Exercise 2.09 and leads to the first equation of motion

$$m(\ddot{s} + \ddot{x}\cos\alpha) = mg\sin\alpha.$$

The second independent virtual displacement can be taken to be an increase of x by δx. In this virtual displacement both m and M move horizontally, so the applied force, gravity, does no work. The incline M has a horizontal acceleration $\ddot{x}$, so the inertial work done on it is $-M\ddot{x}\,\delta x$. The mass m has an acceleration $\mathbf{a}$ as in Exercise 2.09, with a horizontal component $(\ddot{s}\cos\alpha + \ddot{x})$. The inertial work done on m is thus $-m(\ddot{s}\cos\alpha + \ddot{x})\delta x$. D'Alembert's principle gives

$$-M\ddot{x}\,\delta x - m(\ddot{s}\cos\alpha + \ddot{x})\delta x = 0,$$

which simplifies to

$$m(\ddot{s}\cos\alpha + \ddot{x}) = -M\ddot{x}.$$

This second equation of motion can be integrated to yield

$$m\dot{s}\cos\alpha + (m + M)\dot{x} = \text{constant}.$$

In this form the equation states the fact that the horizontal component of the total linear momentum of the system is constant.

CHAPTER III

LAGRANGE'S EQUATIONS

Exercise 3.01

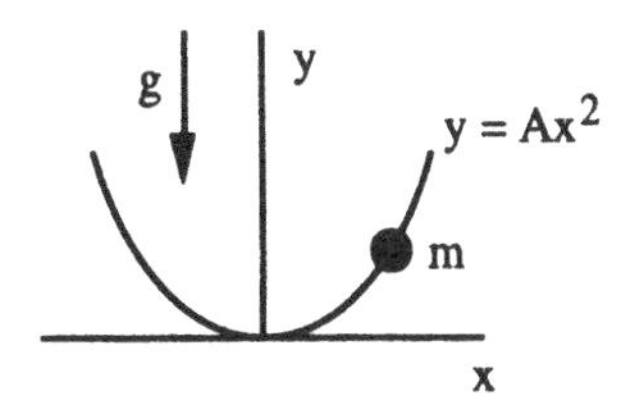

A bead of mass m slides without friction along a wire which has the shape of a parabola $y = Ax^2$ with axis vertical in the earth's gravitational field g.
(a) Find the Lagrangian, taking as generalized coordinate the horizontal displacement x.
(b) Write down Lagrange's equation of motion.

Solution

(a) The kinetic energy of the bead is

$$T = \tfrac{1}{2}m(\dot{x}^2 + \dot{y}^2) = \tfrac{1}{2}m(1 + 4A^2x^2)\dot{x}^2,$$

and the potential energy is

$$V = mgy = mgAx^2.$$

The Lagrangian is thus

$$L(x, \dot{x}) = \tfrac{1}{2}m(1 + 4A^2x^2)\dot{x}^2 - mgAx^2.$$

(b) We have

$$\frac{\partial L}{\partial \dot{x}} = m(1 + 4A^2x^2)\dot{x}, \qquad \frac{\partial L}{\partial x} = \tfrac{1}{2}m(8A^2x)\dot{x}^2 - 2mgAx,$$
$$\frac{d}{dt}\left(\frac{\partial L}{\partial \dot{x}}\right) = m(1 + 4A^2x^2)\ddot{x} + m(8A^2x)\dot{x}^2,$$

so Lagrange's equation of motion is

$$m(1 + 4A^2x^2)\ddot{x} = -m(4A^2x)\dot{x}^2 - 2mgAx.$$

Exercise 3.02

The point of support of a simple plane pendulum moves vertically according to $y = h(t)$, where $h(t)$ is some given function of time.

(a) Find the Lagrangian, taking as generalized coordinate the angle θ the pendulum makes with the vertical.
(b) Write down Lagrange's equation of motion, showing in particular that the pendulum behaves like a simple pendulum in a gravitational field $g + \ddot{h}$.

Solution

(a) The cartesian coordinates of the bob are (Fig. 1)

$$x = \ell\sin\theta, \qquad y = h(t) - \ell\cos\theta.$$

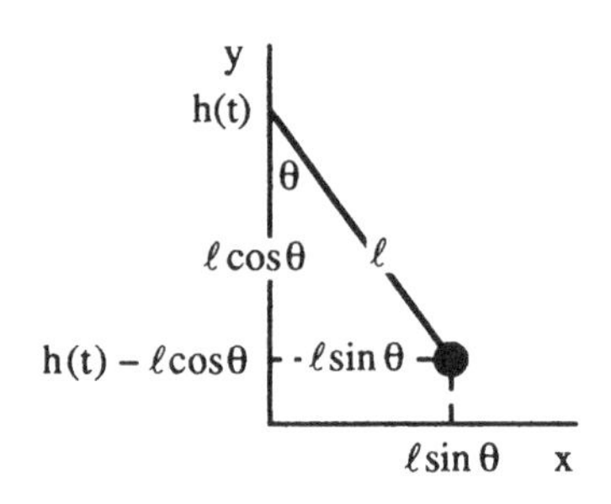

Ex. 3.02, Fig. 1

The cartesian components of the velocity of the bob are thus

$$\dot{x} = \ell\dot{\theta}\cos\theta, \qquad \dot{y} = \dot{h} + \ell\dot{\theta}\sin\theta,$$

and the kinetic energy is

$$T = \tfrac{1}{2}m(\dot{x}^2 + \dot{y}^2) = \tfrac{1}{2}m(\ell^2\dot{\theta}^2 + 2\dot{h}\ell\dot{\theta}\sin\theta + \dot{h}^2).$$

This result can instead be obtained by using the trigonometric cosine law to add the velocity $\dot{h}$ in the vertical direction to the velocity $\ell\dot{\theta}$ in the angular direction (Fig. 2).

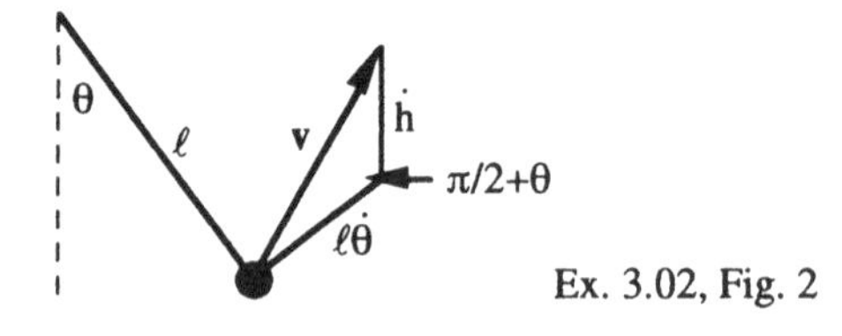

Ex. 3.02, Fig. 2

The potential energy is

$$V = mgy = mg(h - \ell\cos\theta).$$

The Lagrangian is

$$L = T - V = \tfrac{1}{2}m(\ell^2\dot{\theta}^2 + 2\dot{h}\ell\dot{\theta}\sin\theta + \dot{h}^2) - mg(h - \ell\cos\theta).$$

(b) We have

$$\frac{\partial L}{\partial\dot{\theta}} = m(\ell^2\dot{\theta} + \dot{h}\ell\sin\theta), \qquad \frac{\partial L}{\partial\theta} = m\dot{h}\ell\dot{\theta}\cos\theta - mg\ell\sin\theta,$$

$$\frac{d}{dt}\left(\frac{\partial L}{\partial\dot{\theta}}\right) = m(\ell^2\ddot{\theta} + \ddot{h}\ell\sin\theta + \dot{h}\ell\dot{\theta}\cos\theta),$$

so Lagrange's equation is

$$m\ell^2\ddot{\theta} = -m(g + \ddot{h})\ell\sin\theta.$$

This is the same as the equation of motion of a simple pendulum in a gravitational field $g + \ddot{h}$.

Exercise 3.03

A mass m is attached to one end of a light rod of length ℓ. The other end of the rod is pivoted so that the rod can swing in a plane. The pivot rotates in the same plane at angular velocity ω in a circle of radius R. Show that this "pendulum" behaves like a simple pendulum in a gravitational field $g = R\omega^2$ for all values of ℓ and all amplitudes of oscillation.

Solution

The cartesian coordinates of the mass are (Fig. 1)

$$x = R\cos\omega t + \ell\cos(\omega t + \theta), \qquad y = R\sin\omega t + \ell\sin(\omega t + \theta).$$

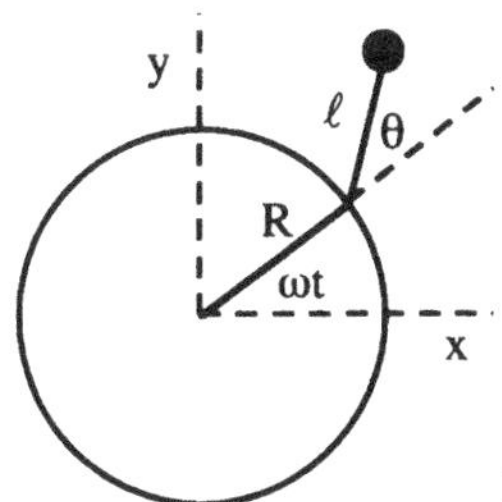

Ex. 3.03, Fig. 1

The cartesian components of the velocity of the mass are thus

$$\dot{x} = -R\omega\sin\omega t - \ell(\omega + \dot{\theta})\sin(\omega t + \theta), \qquad \dot{y} = R\omega\cos\omega t + \ell(\omega + \dot{\theta})\cos(\omega t + \theta),$$

and the kinetic energy is

$$T = \tfrac{1}{2}m(\dot{x}^2 + \dot{y}^2) = \tfrac{1}{2}m\left[R^2\omega^2 + 2R\ell\omega(\omega + \dot{\theta})\cos\theta + \ell^2(\omega + \dot{\theta})^2\right].$$

There is no potential energy, so the Lagrangian is simply T. We have

$$\frac{\partial T}{\partial \dot{\theta}} = mR\ell\omega\cos\theta + m\ell^2(\omega + \dot{\theta}), \qquad \frac{\partial T}{\partial \theta} = -mR\ell\omega(\omega + \dot{\theta})\sin\theta,$$

$$\frac{d}{dt}\left(\frac{\partial T}{\partial \dot{\theta}}\right) = -mR\ell\omega\dot{\theta}\sin\theta + m\ell^2\ddot{\theta},$$

so Lagrange's equation is

$$m\ell^2\ddot{\theta} = -mR\omega^2\ell\sin\theta.$$

This is the same as the equation of motion of a simple pendulum in a gravitational field $R\omega^2$.

Exercise 3.04

A pendulum is formed by suspending a mass m from the ceiling, using a spring of unstretched length ℓ_0 and spring constant k.
(a) Choose, and show on a diagram, appropriate generalized coordinates, assuming that the pendulum moves in a fixed vertical plane.
(b) Set up the Lagrangian using your generalized coordinates.
(c) Write down the explicit Lagrange's equations of motion for your generalized coordinates.

Solution

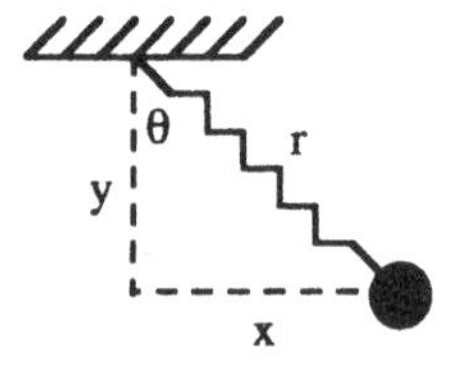

Ex. 3.04, Fig. 1

If we choose as generalized coordinates the cartesian coordinates (x,y) with origin at the point of suspension and with x horizontal and y vertically down (Fig. 1), the Lagrangian is

$$L = \tfrac{1}{2}m(\dot{x}^2 + \dot{y}^2) + mgy - \tfrac{1}{2}k(\sqrt{x^2 + y^2} - \ell_0)^2.$$

The second term is (minus) the gravitational potential energy and the third term is (minus) the potential energy due to the stretch of the spring. Lagrange's equations are

$$m\ddot{x} = -kx(1 - \ell_0/r), \qquad m\ddot{y} = mg - ky(1 - \ell_0/r),$$

with $r = \sqrt{x^2 + y^2}$.

If, instead, we choose as generalized coordinates the polar coordinates (r,θ) with r the distance of the mass from the point of suspension and θ the angle the spring makes with the vertical (Fig. 1), the Lagrangian is

$$L = \tfrac{1}{2}m(\dot{r}^2 + r^2\dot{\theta}^2) + mgr\cos\theta - \tfrac{1}{2}k(r - \ell_0)^2.$$

Lagrange's equations are then

$$m\ddot{r} = mr\dot{\theta}^2 + mg\cos\theta - k(r - \ell_0),$$
$$d(mr^2\dot{\theta})/dt = mr^2\ddot{\theta} + 2mr\dot{r}\dot{\theta} = -mgr\sin\theta.$$

Exercise 3.05

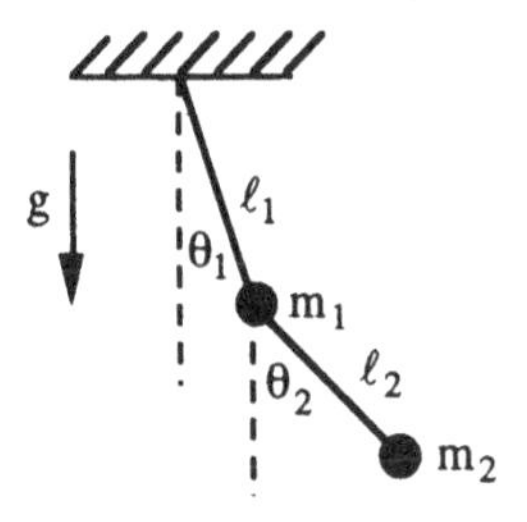

A double plane pendulum consists of two simple pendulums, with one pendulum suspended from the bob of the other. The "upper" pendulum has mass m_1 and length ℓ_1, the "lower" pendulum has mass m_2 and length ℓ_2, and both pendulums move in the same vertical plane.

(a) Find the Lagrangian, using as generalized coordinates the angles θ_1 and θ_2 the pendulums make with the vertical.

(b) Write down Lagrange's equations of motion.

Solution

(a) The cartesian coordinates of m_1 are

$$x_1 = \ell_1 \sin\theta_1, \qquad y_1 = \ell_1 \cos\theta_1,$$

so the cartesian components of the velocity of m_1 are

$$\dot{x}_1 = \ell_1\dot{\theta}_1 \cos\theta_1, \qquad \dot{y}_1 = -\ell_1\dot{\theta}_1 \sin\theta_1.$$

The kinetic energy of m_1 is

$$T_1 = \tfrac{1}{2}m_1(\dot{x}_1^2 + \dot{y}_1^2) = \tfrac{1}{2}m_1\ell_1^2\dot{\theta}_1^2.$$

The cartesian coordinates of m_2 are

$$x_2 = \ell_1 \sin\theta_1 + \ell_2 \sin\theta_2, \qquad y_2 = \ell_1 \cos\theta_1 + \ell_2 \cos\theta_2,$$

so the cartesian components of the velocity of m_2 are

$$\dot{x}_2 = \ell_1\dot{\theta}_1 \cos\theta_1 + \ell_2\dot{\theta}_2 \cos\theta_2, \qquad \dot{y}_1 = -\ell_1\dot{\theta}_1 \sin\theta_1 - \ell_2\dot{\theta}_2 \sin\theta_2.$$

The kinetic energy of m_2 is

$$T_2 = \tfrac{1}{2}m_2(\dot{x}_2^2 + \dot{y}_2^2) = \tfrac{1}{2}m_2(\ell_1^2\dot{\theta}_1^2 + 2\ell_1\ell_2\dot{\theta}_1\dot{\theta}_2 \cos(\theta_2 - \theta_1) + \ell_2^2\dot{\theta}_2^2).$$

The total potential energy is

$$V = -m_1 g\ell_1 \cos\theta_1 - m_2 g(\ell_1 \cos\theta_1 + \ell_2 \cos\theta_2).$$

The Lagrangian is

$$\begin{aligned} L = {} & \tfrac{1}{2}m_1\ell_1^2\dot{\theta}_1^2 + \tfrac{1}{2}m_2(\ell_1^2\dot{\theta}_1^2 + 2\ell_1\ell_2\dot{\theta}_1\dot{\theta}_2 \cos(\theta_2 - \theta_1) + \ell_2^2\dot{\theta}_2^2) \\ & + m_1 g\ell_1 \cos\theta_1 + m_2 g(\ell_1 \cos\theta_1 + \ell_2 \cos\theta_2)\,. \end{aligned}$$

(b) We have

$$\frac{\partial L}{\partial\dot{\theta}_1} = (m_1 + m_2)\ell_1^2\dot{\theta}_1 + m_2\ell_1\ell_2\dot{\theta}_2 \cos(\theta_2 - \theta_1),$$

$$\frac{d}{dt}\left(\frac{\partial L}{\partial\dot{\theta}_1}\right) = (m_1 + m_2)\ell_1^2\ddot{\theta}_1 + m_2\ell_1\ell_2\ddot{\theta}_2 \cos(\theta_2 - \theta_1) - m_2\ell_1\ell_2\dot{\theta}_2(\dot{\theta}_2 - \dot{\theta}_1)\sin(\theta_2 - \theta_1),$$

$$\frac{\partial L}{\partial\theta_1} = m_2\ell_1\ell_2\dot{\theta}_1\dot{\theta}_2 \sin(\theta_2 - \theta_1) - (m_1 + m_2)g\ell_1 \sin\theta_1,$$

and

$$\frac{\partial L}{\partial \dot{\theta}_2} = m_2\ell_2^2\dot{\theta}_2 + m_2\ell_1\ell_2\dot{\theta}_1\cos(\theta_2 - \theta_1),$$

$$\frac{d}{dt}\left(\frac{\partial L}{\partial \dot{\theta}_2}\right) = m_2\ell_2^2\ddot{\theta}_2 + m_2\ell_1\ell_2\ddot{\theta}_1\cos(\theta_2 - \theta_1) - m_2\ell_1\ell_2\dot{\theta}_1(\dot{\theta}_2 - \dot{\theta}_1)\sin(\theta_2 - \theta_1),$$

$$\frac{\partial L}{\partial \theta_2} = -m_2\ell_1\ell_2\dot{\theta}_1\dot{\theta}_2\sin(\theta_2 - \theta_1) - m_2 g\ell_2\sin\theta_2.$$

Lagrange's equations are

$$(m_1 + m_2)\ell_1^2\ddot{\theta}_1 + m_2\ell_1\ell_2\ddot{\theta}_2\cos(\theta_2 - \theta_1) - m_2\ell_1\ell_2\dot{\theta}_2^2\sin(\theta_2 - \theta_1) = -(m_1 + m_2)g\ell_1\sin\theta_1,$$

$$m_2\ell_2^2\ddot{\theta}_2 + m_2\ell_1\ell_2\ddot{\theta}_1\cos(\theta_2 - \theta_1) + m_2\ell_1\ell_2\dot{\theta}_1^2\sin(\theta_2 - \theta_1) = -m_2\ell_2 g\sin\theta_2.$$

In the small angle approximation these become

$$(m_1 + m_2)\ell_1^2\ddot{\theta}_1 + m_2\ell_1\ell_2\ddot{\theta}_2 \approx -(m_1 + m_2)g\ell_1\theta_1,$$

$$m_2\ell_2^2\ddot{\theta}_2 + m_2\ell_1\ell_2\ddot{\theta}_1 \approx -m_2\ell_2 g\theta_2,$$

and are the equations of motion of two linearly coupled simple harmonic oscillators.

Exercise 3.06

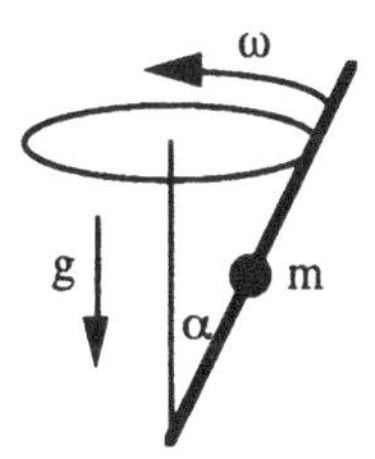

A bead of mass m slides on a long straight wire which makes an angle α with, and rotates with constant angular velocity ω about, the upward vertical. Gravity g acts vertically downwards.

(a) Choose an appropriate generalized coordinate and find the Lagrangian.

(b) Write down the explicit Lagrange's equation of motion.

Solution

(a) Choose as generalized coordinate the distance r of the bead along the wire from the axis of rotation. The kinetic energy of the bead is then

$$T = \tfrac{1}{2}m(\dot{r}^2 + \omega^2 r^2 \sin^2\alpha),$$

and the gravitational potential energy is

$$V = mgr\cos\alpha.$$

The Lagrangian is

$$L = \tfrac{1}{2}m(\dot{r}^2 + \omega^2 r^2 \sin^2\alpha) - mgr\cos\alpha.$$

(b) Lagrange's equation is

$$m\ddot{r} = m\omega^2 r\sin^2\alpha - mg\cos\alpha.$$

This is the same as the equation for one-dimensional motion in an effective potential

$$V_{eff} = -\tfrac{1}{2}m\omega^2 r^2 \sin^2\alpha + mgr\cos\alpha.$$

This potential has a maximum at $r = \dfrac{g}{\omega^2}\dfrac{\cos\alpha}{\sin^2\alpha}$, corresponding to an unstable equilibrium point.

Exercise 3.07

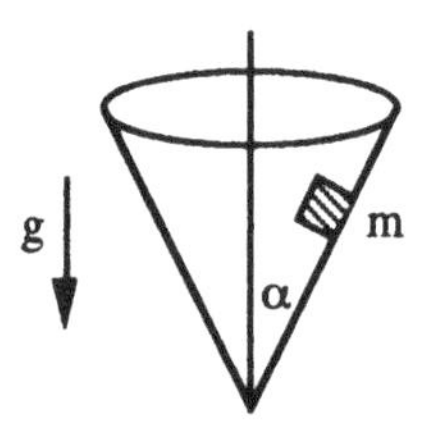

A particle of mass m slides on the inner surface of a cone of half angle α. The axis of the cone is vertical with vertex downward. Gravity g acts vertically downwards.
(a) Choose and show on a diagram suitable generalized coordinates, and find the Lagrangian.
(b) Write down the explicit equations of motion for your generalized coordinates.

Solution

(a) Choose as generalized coordinates the distance r of the mass from the apex of the cone together with the angle ϕ measured in a horizontal circle around the axis of the cone. The coordinates r and ϕ are simply two of the usual spherical polar coordinates. The third spherical polar coordinate θ is, in this problem, fixed at $\theta = \alpha$. The Lagrangian is

$$L = \tfrac{1}{2}m(\dot{r}^2 + r^2\dot{\phi}^2 \sin^2\alpha) - mgr\cos\alpha .$$

(b) Lagrange's equations are

$$m\ddot{r} = mr\dot{\phi}^2 \sin^2\alpha - mg\cos\alpha , \qquad d(mr^2\dot{\phi}\sin^2\alpha)/dt = 0 .$$

The second of these equations yields $mr^2\dot{\phi}\sin^2\alpha = L_z$, where L_z is a constant which can be identified as the angular momentum of the particle in the vertical (z-) direction. Substituting this into the first equation, we obtain

$$m\frac{d^2r}{dt^2} = \frac{L_z^2}{mr^3\sin^2\alpha} - mg\cos\alpha = -\frac{d}{dr}\left(\frac{L_z^2}{2mr^2\sin^2\alpha} + mgr\cos\alpha\right).$$

This is the same as the equation for one-dimensional motion in an effective potential

$$V_{eff} = \frac{L_z^2}{2mr^2\sin^2\alpha} + mgr\cos\alpha .$$

V_{eff} has a minimum at $r^3 = \dfrac{L_z^2}{m^2 g\sin^2\alpha\cos\alpha}$, corresponding to a stable horizontal circular orbit.

Exercise 3.08

Using spherical polar coordinates (r,θ,ϕ) defined by

$$x = r\sin\theta\cos\phi \quad y = r\sin\theta\sin\phi \quad z = r\cos\theta,$$

write down the Lagrangian and find the explicit Lagrange's equations of motion for a particle of mass m moving in a central potential $V(r)$.

Solution

The components of the velocity of a particle in spherical polar coordinates are $\dot{r}$ in the r-direction, $r\dot{\theta}$ in the θ-direction, and $r\dot{\phi}\sin\theta$ in the ϕ-direction. Since these are mutually perpendicular, the speed v of the particle is given by

$$v^2 = \dot{r}^2 + r^2\dot{\theta}^2 + r^2\dot{\phi}^2\sin^2\theta,$$

and its kinetic energy by

$$T = \tfrac{1}{2}mv^2 = \tfrac{1}{2}m(\dot{r}^2 + r^2\dot{\theta}^2 + r^2\dot{\phi}^2\sin^2\theta).$$

Alternatively, this can be obtained by transforming the expression

$$T = \tfrac{1}{2}m(\dot{x}^2 + \dot{y}^2 + \dot{z}^2)$$

for the kinetic energy in cartesian coordinates, by setting

$$\begin{aligned} x &= r\sin\theta\cos\phi, & \dot{x} &= \dot{r}\sin\theta\cos\phi + r\dot{\theta}\cos\theta\cos\phi - r\dot{\phi}\sin\theta\sin\phi, \\ y &= r\sin\theta\sin\phi, & \dot{y} &= \dot{r}\sin\theta\sin\phi + r\dot{\theta}\cos\theta\sin\phi + r\dot{\phi}\sin\theta\cos\phi, \\ z &= r\cos\theta, & \dot{z} &= \dot{r}\cos\theta - r\dot{\theta}\sin\theta. \end{aligned}$$

If the particle moves in a central potential $V(r)$, the Lagrangian is

$$L = T - V = \tfrac{1}{2}m(\dot{r}^2 + r^2\dot{\theta}^2 + r^2\dot{\phi}^2\sin^2\theta) - V(r).$$

We have

$$\begin{aligned} \frac{\partial L}{\partial \dot{r}} &= m\dot{r}, & \frac{\partial L}{\partial \dot{\theta}} &= mr^2\dot{\theta}, & \frac{\partial L}{\partial \dot{\phi}} &= mr^2\dot{\phi}\sin^2\theta, \\ \frac{\partial L}{\partial r} &= mr(\dot{\theta}^2 + \dot{\phi}^2\sin^2\theta) - \frac{dV}{dr}, & \frac{\partial L}{\partial \theta} &= mr^2\dot{\phi}^2\sin\theta\cos\theta, & \frac{\partial L}{\partial \phi} &= 0, \end{aligned}$$

so Lagrange's equations in spherical polar coordinates for a particle moving in a central potential are

$$\begin{aligned} &\frac{d}{dt}(m\dot{r}) = mr(\dot{\theta}^2 + \dot{\phi}^2\sin^2\theta) - \frac{dV(r)}{dr}, \\ &\frac{d}{dt}(mr^2\dot{\theta}) = mr^2\dot{\phi}^2\sin\theta\cos\theta, \\ &\frac{d}{dt}(mr^2\dot{\phi}\sin^2\theta) = 0. \end{aligned}$$

Exercise 3.09

For some problems paraboloidal coordinates (ξ,η,ϕ) defined by

$$x=\xi\eta\cos\phi \quad y=\xi\eta\sin\phi \quad z=\tfrac{1}{2}(\xi^2-\eta^2)$$

turn out to be convenient.
(a) Show that the surfaces $\xi=$ const. or $\eta=$ const. are paraboloids of revolution about the z-axis with focus at the origin and semi-latus-rectum ξ^2 or η^2.
(b) Express the kinetic energy of a particle of mass m in terms of paraboloidal coordinates and their first time derivatives.
(Ans. $T=\frac{1}{2}m\left(\xi^2+\eta^2\right)\left(\dot{\xi}^2+\dot{\eta}^2\right)+\frac{1}{2}m\xi^2\eta^2\dot{\phi}^2$)

Solution

(a) Paraboloidal coordinates (ξ,η,ϕ) are defined by

$$x=\xi\eta\cos\phi, \quad y=\xi\eta\sin\phi, \quad z=\tfrac{1}{2}(\xi^2-\eta^2).$$

Note also that the cylindrical coordinate ρ (the distance from the z-axis) is

$$\rho=\sqrt{x^2+y^2}=\xi\eta$$

and that the spherical polar coordinate r (the distance from the origin) is

$$r=\sqrt{x^2+y^2+z^2}=\tfrac{1}{2}(\xi^2+\eta^2).$$

We have

$$r+z=\xi^2 \quad \text{and} \quad r-z=\eta^2.$$

Setting $r=\sqrt{\rho^2+z^2}$ in these and solving for z, we find

$$z=\frac{\xi^2}{2}-\frac{\rho^2}{2\xi^2} \quad \text{and} \quad z=-\frac{\eta^2}{2}+\frac{\rho^2}{2\eta^2}.$$

For fixed ξ or η these are the equations of paraboloids of revolution about the z-axis; the fixed-ξ ones open in the negative z-direction and the fixed-η ones open in the positive z-direction (Fig. 1).

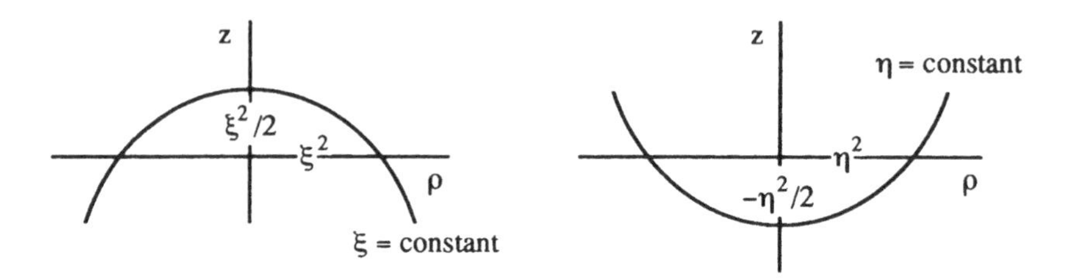

Ex. 3.09, Fig. 1

(b) The element of distance in paraboloidal coordinates is obtained most easily from that in cylindrical coordinates,

$$(ds)^2 = (d\rho)^2 + \rho^2(d\phi)^2 + (dz)^2,$$

by setting

$$d\rho = \xi d\eta + \eta d\xi \quad \text{and} \quad dz = \xi d\xi - \eta d\eta$$

together with $\rho = \xi\eta$. This gives

$$(ds)^2 = (\xi^2 + \eta^2)\left((d\xi)^2 + (d\eta)^2\right) + \xi^2\eta^2(d\phi)^2.$$

The kinetic energy of a particle of mass m is then

$$T = \frac{1}{2}m\left(\frac{ds}{dt}\right)^2 = \frac{1}{2}m\left[(\xi^2 + \eta^2)\left(\dot{\xi}^2 + \dot{\eta}^2\right) + \xi^2\eta^2\dot{\phi}^2\right]$$

in paraboloidal coordinates.

Exercise 3.10

The motion of a particle of mass m is given by Lagrange's equations with Lagrangian

$$L = \exp(\alpha t/m)(T - V)$$

where α is a constant, $T = \frac{1}{2}m(\dot{x}^2 + \dot{y}^2 + \dot{z}^2)$ is the kinetic energy, and $V = V(x, y, z)$ is the potential energy. Write down the equations of motion and interpret.

Solution

We have

$$\frac{\partial L}{\partial \dot{x}} = e^{\alpha t/m} m\dot{x}, \qquad \frac{d}{dt}\left(\frac{\partial L}{\partial \dot{x}}\right) = e^{\alpha t/m}(m\ddot{x} + \alpha\dot{x}),$$
$$\frac{\partial L}{\partial x} = e^{\alpha t/m}\left(-\partial V/\partial x\right),$$

so Lagrange's equation for x is

$$m\ddot{x} = -\alpha\dot{x} - \partial V/\partial x$$

with similar equations for y and z. These equations can be combined into the single vector equation

$$m\ddot{\mathbf{r}} = -\alpha\dot{\mathbf{r}} - \nabla V.$$

This is the equation of motion for a particle of mass m which is acted on by a conservative force $-\nabla V$ together with a frictional force $-\alpha\dot{\mathbf{r}}$ which is proportional to the velocity of the particle.

Exercise 3.11

A system with two degrees of freedom (x, y) is described by a Lagrangian

$$L = \tfrac{1}{2}m(a\dot{x}^2 + 2b\dot{x}\dot{y} + c\dot{y}^2) - \tfrac{1}{2}k(ax^2 + 2bxy + cy^2)$$

where a, b, and c are constants, with $b^2 \neq ac$. Write down Lagrange's equations of motion and thereby identify the system. Consider in particular the cases $a = c = 0, b \neq 0$ and $a = -c, b = 0$.

Solution

We have

$$\frac{\partial L}{\partial \dot{x}} = m(a\dot{x} + b\dot{y}), \qquad \frac{\partial L}{\partial x} = -k(ax + by),$$
$$\frac{\partial L}{\partial \dot{y}} = m(b\dot{x} + c\dot{y}), \qquad \frac{\partial L}{\partial y} = -k(bx + cy),$$

so Lagrange's equations are

$$m(a\ddot{x} + b\ddot{y}) = -k(ax + by),$$
$$m(b\ddot{x} + c\ddot{y}) = -k(bx + cy).$$

These can be written in matrix form

$$m\begin{bmatrix} a & b \\ b & c \end{bmatrix}\begin{bmatrix} \ddot{x} \\ \ddot{y} \end{bmatrix} = -k\begin{bmatrix} a & b \\ b & c \end{bmatrix}\begin{bmatrix} x \\ y \end{bmatrix}.$$

If we multiply both sides of this equation by the matrix reciprocal to $\begin{bmatrix} a & b \\ b & c \end{bmatrix}$ (note that the reciprocal exists if the determinant $ac - b^2$ of the matrix is non-zero), we obtain

$$m\ddot{x} = -kx,$$
$$m\ddot{y} = -ky.$$

These are the equations of motion of a two-dimensional harmonic oscillator (a particle of mass m constrained to move in the xy-plane and pulled towards the origin by a spring of zero unstretched length and spring constant k). The usual Lagrangian

$$L_0 = \tfrac{1}{2}m(\dot{x}^2 + \dot{y}^2) - \tfrac{1}{2}k(x^2 + y^2)$$

(kinetic energy minus potential energy) for this system is obtained by setting $a = c = 1$ and $b = 0$. It is clear from this example, however, that it is sometimes possible to find other Lagrangians which lead to the same equations of motion. Thus, for example, if we take $a = c = 0$ and $b = 1$ we get an equivalent Lagrangian

$$L_1 = m\dot{x}\dot{y} - kxy,$$

whereas if we take $a = -c = 1$ and $b = 0$ we get an equivalent Lagrangian

$$L_2 = \tfrac{1}{2}m(\dot{x}^2 - \dot{y}^2) - \tfrac{1}{2}k(x^2 - y^2).$$

Exercise 3.12

The Lagrangian for two particles of masses m_1 and m_2 and coordinates $\mathbf{r}_1$ and $\mathbf{r}_2$, interacting via a potential $V(\mathbf{r}_1 - \mathbf{r}_2)$, is

$$L = \tfrac{1}{2}m_1|\dot{\mathbf{r}}_1|^2 + \tfrac{1}{2}m_2|\dot{\mathbf{r}}_2|^2 - V(\mathbf{r}_1 - \mathbf{r}_2).$$

(a) Rewrite the Lagrangian in terms of the center of mass coordinates $\mathbf{R} = \dfrac{m_1\mathbf{r}_1 + m_2\mathbf{r}_2}{m_1 + m_2}$ and relative coordinates $\mathbf{r} = \mathbf{r}_1 - \mathbf{r}_2$.
(b) Use Lagrange's equations to show that the center of mass and relative motions separate, the center of mass moving with constant velocity, and the relative motion being like that of a particle of reduced mass $\dfrac{m_1 m_2}{m_1 + m_2}$ in a potential $V(\mathbf{r})$.

Solution

(a) The coordinates $\mathbf{r}_1$ and $\mathbf{r}_2$ of the particles, in terms of the center of mass coordinates $\mathbf{R}$ and the relative coordinates $\mathbf{r}$, are (Fig. 1)

$$\mathbf{r}_1 = \mathbf{R} + \frac{m_2}{m_1 + m_2}\mathbf{r} \quad \text{and} \quad \mathbf{r}_2 = \mathbf{R} - \frac{m_1}{m_1 + m_2}\mathbf{r}.$$

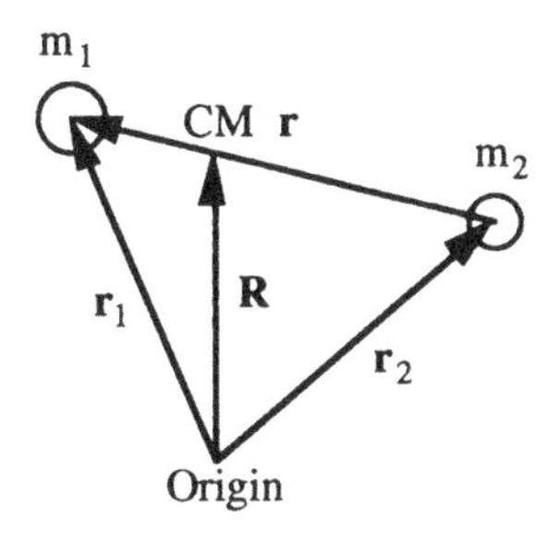

Ex. 3.12, Fig. 1

The Lagrangian is

$$L = \frac{1}{2}m_1\left|\dot{\mathbf{R}} + \frac{m_2}{m_1 + m_2}\dot{\mathbf{r}}\right|^2 + \frac{1}{2}m_2\left|\dot{\mathbf{R}} - \frac{m_1}{m_1 + m_2}\dot{\mathbf{r}}\right|^2 - V(\mathbf{r})$$
$$= \frac{1}{2}(m_1 + m_2)|\dot{\mathbf{R}}|^2 + \frac{1}{2}\frac{m_1 m_2}{m_1 + m_2}|\dot{\mathbf{r}}|^2 - V(\mathbf{r}).$$

(b) Lagrange's equations for $\mathbf{R}$ and $\mathbf{r}$ are

$$(m_1 + m_2)\ddot{\mathbf{R}} = 0 \quad \text{and} \quad \frac{m_1 m_2}{m_1 + m_2}\ddot{\mathbf{r}} = -\frac{\partial V(\mathbf{r})}{\partial \mathbf{r}}.$$

The first of these shows that the center of mass moves with constant velocity, while the second shows that the relative motion is the same as that of a particle of "reduced mass" $\frac{m_1 m_2}{m_1 + m_2}$ in a potential $V(\mathbf{r})$.

Exercise 3.13*

Consider the motion of a free particle, with Lagrangian

$$L = \tfrac{1}{2}m(\dot{x}^2 + \dot{y}^2 + \dot{z}^2),$$

as viewed from a rotating coordinate system

$$x' = x\cos\theta + y\sin\theta, \quad y' = -x\sin\theta + y\cos\theta, \quad z' = z$$

where the angle $\theta = \theta(t)$ is some given function of time.
(a) Show that in terms of these coordinates the Lagrangian takes the form

$$L' = \tfrac{1}{2}m\left[(\dot{x}'^2 + \dot{y}'^2 + \dot{z}'^2) + 2\omega(x'\dot{y}' - y'\dot{x}') + \omega^2(x'^2 + y'^2)\right]$$

where $\omega = d\theta/dt$ is the angular velocity.
(b) Write down Lagrange's equations of motion, and show that they look like those for a particle which is acted on by a "force." The part of the "force" proportional to ω is called the Coriolis force, that proportional to ω^2 is called the centrifugal force, and that proportional to $d\omega/dt$ is called the Euler force. Identify the components of these "forces."

Solution

(a) To determine the Lagrangian to be used in a rotating cartesian frame, we transform variables, setting in L

$$\begin{aligned} x &= x'\cos\theta - y'\sin\theta, & \dot{x} &= \dot{x}'\cos\theta - \dot{y}'\sin\theta - \omega x'\sin\theta - \omega y'\cos\theta, \\ y &= x'\sin\theta + y'\cos\theta, & \dot{y} &= \dot{x}'\sin\theta + \dot{y}'\cos\theta + \omega x'\cos\theta - \omega y'\sin\theta, \\ z &= z', & \dot{z} &= \dot{z}', \end{aligned}$$

where $\omega = d\theta/dt$ is the angular velocity of the rotating frame. We thus find

$$\begin{aligned} L' &= \tfrac{1}{2}m\Big[(\dot{x}'\cos\theta - \dot{y}'\sin\theta - \omega x'\sin\theta - \omega y'\cos\theta)^2 + \\ &\qquad + (\dot{x}'\sin\theta + \dot{y}'\cos\theta + \omega x'\cos\theta - \omega y'\sin\theta)^2 + (\dot{z}')^2\Big] \\ &= \tfrac{1}{2}m\left[\dot{x}'^2 + \dot{y}'^2 + \dot{z}'^2 + 2\omega(x'\dot{y}' - y'\dot{x}') + \omega^2(x'^2 + y'^2)\right]. \end{aligned}$$

(b) We have

$$\frac{\partial L'}{\partial \dot{x}'} = m\dot{x}' - m\omega y', \quad \frac{d}{dt}\left(\frac{\partial L'}{\partial \dot{x}'}\right) = m\ddot{x}' - m\omega\dot{y}' - m\frac{d\omega}{dt}y', \quad \frac{\partial L'}{\partial x'} = m\omega\dot{y}' + m\omega^2 x',$$

$$\frac{\partial L'}{\partial \dot{y}'} = m\dot{y}' + m\omega x', \quad \frac{d}{dt}\left(\frac{\partial L'}{\partial \dot{y}'}\right) = m\ddot{y}' + m\omega\dot{x}' + m\frac{d\omega}{dt}x', \quad \frac{\partial L'}{\partial x'} = -m\omega\dot{x}' + m\omega^2 y',$$

$$\frac{\partial L'}{\partial \dot{z}'} = m\dot{z}', \qquad \frac{\partial L'}{\partial \dot{z}'} = m\dot{z}', \qquad \frac{\partial L'}{\partial z'} = 0,$$

so Lagrange's equations in the rotating frame are

$$\begin{aligned} m\ddot{x}' &= 2m\omega\dot{y}' + m\omega^2 x' + m(d\omega/dt)y', \\ m\ddot{y}' &= -2m\omega\dot{x}' + m\omega^2 y' - m(d\omega/dt)x', \\ m\ddot{z}' &= 0. \end{aligned}$$

The right-hand sides of these equations are the components of the "inertial force" which observers in the rotating frame say acts on the particle. The terms proportional to ω are the "Coriolis force"

$$\mathbf{F}'(\text{Coriolis}) = (2m\omega\dot{y}', -2m\omega\dot{x}', 0) = -2m\boldsymbol{\omega} \times \dot{\mathbf{r}}';$$

the terms proportional to ω^2 are the "centrifugal force"

$$\mathbf{F}'(\text{centrifugal}) = (m\omega^2 x', m\omega^2 y', 0) = -m\boldsymbol{\omega} \times (\boldsymbol{\omega} \times \mathbf{r}');$$

and the terms proportional to $d\omega/dt$ are the "Euler force"

$$\mathbf{F}'(\text{Euler}) = \left(m\frac{d\omega}{dt}y', -m\frac{d\omega}{dt}x', 0 \right) = -m\frac{d\boldsymbol{\omega}}{dt} \times \mathbf{r}'.$$

The Euler force is zero if the rotation is uniform.

Exercise 3.14

(a) Write down the equations of motion resulting from a Lagrangian

$$L = \tfrac{1}{2}m(\dot{x}^2 + \dot{y}^2 + \dot{z}^2) - V(r) + (eB/2c)(x\dot{y} - y\dot{x}),$$

and show that they are those for a particle of mass m and charge e moving in a central potential $V(r)$ together with a uniform magnetic field B which points in the z-direction.
(b) Suppose, instead of the inertial cartesian coordinate system (x, y, z), we use a rotating system (x', y', z') with

$$x' = x\cos\omega t + y\sin\omega t, \qquad y' = -x\sin\omega t + y\cos\omega t, \qquad z' = z.$$

Change variables, obtaining the above Lagrangian in terms of (x', y', z') and their first time derivatives. Show that we can eliminate the term linear in B by an appropriate choice of ω (this is Larmor's theorem: the effect of a weak magnetic field on a system is to induce a uniform rotation at frequency ω_L, the Larmor frequency).

Solution

(a) We have

$$\frac{\partial L}{\partial \dot{x}} = m\dot{x} - \frac{eB}{2c}y, \qquad \frac{\partial L}{\partial \dot{y}} = m\dot{y} + \frac{eB}{2c}x, \qquad \frac{\partial L}{\partial \dot{z}} = m\dot{z},$$

$$\frac{d}{dt}\left(\frac{\partial L}{\partial \dot{x}}\right) = m\ddot{x} - \frac{eB}{2c}\dot{y}, \qquad \frac{d}{dt}\left(\frac{\partial L}{\partial \dot{y}}\right) = m\ddot{y} + \frac{eB}{2c}\dot{x}, \qquad \frac{d}{dt}\left(\frac{\partial L}{\partial \dot{z}}\right) = m\ddot{z},$$

$$\frac{\partial L}{\partial x} = -\frac{\partial V}{\partial x} + \frac{eB}{2c}\dot{y}, \qquad \frac{\partial L}{\partial y} = -\frac{\partial V}{\partial y} - \frac{eB}{2c}\dot{x}, \qquad \frac{\partial L}{\partial z} = -\frac{\partial V}{\partial z},$$

so Lagrange's equations are

$$m\ddot{x} = -\frac{\partial V}{\partial x} + \frac{eB}{c}\dot{y}, \quad m\ddot{y} = -\frac{\partial V}{\partial y} - \frac{eB}{c}\dot{x}, \quad m\ddot{z} = -\frac{\partial V}{\partial z}.$$

These can be combined into the single vector equation

$$m\ddot{\mathbf{r}} = -\nabla V + (e/c)\dot{\mathbf{r}} \times \mathbf{B}$$

(where $\mathbf{B} = B\mathbf{k}$) and are the equations of motion, in an inertial frame, of a particle of mass m and charge e in a potential V together with a uniform magnetic field $\mathbf{B}$ in the z-direction.

(b) The Lagrangian in a rotating frame can be obtained by transforming the variables as in Exercise 3-13. The first term in L is the same as the Lagrangian in Exercise 3-13 and so transforms the same; the second term V(r) is invariant; and the third transforms as

$$x\dot{y} - y\dot{x} = x'\dot{y}' - y'\dot{x}' + \omega(x'^2 + y'^2).$$

The Lagrangian in the rotating frame is thus

$$L' = \frac{1}{2}m(\dot{x}'^2 + \dot{y}'^2 + \dot{z}'^2) - V(r)$$
$$+ m\left(\omega + \frac{eB}{2mc}\right)(x'\dot{y}' - y'\dot{x}') + \frac{1}{2}m\left(\omega^2 + \frac{eB}{mc}\omega\right)(x'^2 + y'^2).$$

If we take $\omega = \omega_L = -\frac{eB}{2mc}$, the "Larmor frequency," the third term in L' is eliminated and the Lagrangian reduces to

$$L' = \frac{1}{2}m(\dot{x}'^2 + \dot{y}'^2 + \dot{z}'^2) - V(r) - \frac{1}{2}m\omega_L^2(x'^2 + y'^2).$$

It now contains only second order terms in the magnetic field. These are small if the magnetic field is weak.

Exercise 3.15

Show that the equations of motion of an electric charge e interacting with a magnet of moment $\mathbf{m}$ can be obtained from a Lagrangian

$$L = \tfrac{1}{2}M_e v_e^2 + \tfrac{1}{2}M_m v_m^2 + (e/c)(\mathbf{v}_e - \mathbf{v}_m)\cdot\mathbf{A}(\mathbf{r}_e - \mathbf{r}_m),$$

where

$$\mathbf{A}(\mathbf{r}_e - \mathbf{r}_m) = \frac{\mathbf{m}\times(\mathbf{r}_e - \mathbf{r}_m)}{|\mathbf{r}_e - \mathbf{r}_m|^3}$$

is the vector potential at the charge due to the magnet.
(Y. Aharonov and A. Casher, "Topological Quantum Effects for Neutral Particles," Phys. Rev. Lett. **53**, 319-321 (1984)).

Solution

First consider the charge. We have

$$\frac{\partial L}{\partial \mathbf{v}_e} = M_e\mathbf{v}_e + \frac{e}{c}\mathbf{A},$$

$$\frac{d}{dt}\left(\frac{\partial L}{\partial \mathbf{v}_e}\right) = M_e\frac{d\mathbf{v}_e}{dt} + \frac{e}{c}\frac{d\mathbf{A}}{dt} = M_e\frac{d\mathbf{v}_e}{dt} + \frac{e}{c}\left[\frac{d\mathbf{m}}{dt}\times\frac{\mathbf{r}_e - \mathbf{r}_m}{|\mathbf{r}_e - \mathbf{r}_m|^3} + (\mathbf{v}_e - \mathbf{v}_m)\cdot\nabla_e\mathbf{A}\right],$$

$$\frac{\partial L}{\partial \mathbf{r}_e} = \frac{e}{c}\nabla_e[(\mathbf{v}_e - \mathbf{v}_m)\cdot\mathbf{A}] = \frac{e}{c}[(\mathbf{v}_e - \mathbf{v}_m)\cdot\nabla_e\mathbf{A} + (\mathbf{v}_e - \mathbf{v}_m)\times(\nabla_e\times\mathbf{A})].$$

Lagrange's equations for the charge are

$$M_e\frac{d\mathbf{v}_e}{dt} = e\left[-\frac{1}{c}\frac{d\mathbf{m}}{dt}\times\frac{\mathbf{r}_e - \mathbf{r}_m}{|\mathbf{r}_e - \mathbf{r}_m|^3} - \frac{1}{c}\mathbf{v}_m\times(\nabla_e\times\mathbf{A})\right] + \frac{e}{c}\mathbf{v}_e\times(\nabla_e\times\mathbf{A}).$$

Now

$$\mathbf{B}_m = \nabla_e\times\mathbf{A}$$

is the magnetic field due to the magnet at the location of the charge. The term in square brackets,

$$\mathbf{E}_m = -\frac{1}{c}\frac{d\mathbf{m}}{dt}\times\frac{\mathbf{r}_e - \mathbf{r}_m}{|\mathbf{r}_e - \mathbf{r}_m|^3} - \frac{1}{c}\mathbf{v}_m\times\mathbf{B}_m,$$

is the electric field due to the magnet at the location of the charge. It consists of two parts: the first is the electric field due to the time-dependence of the magnetic moment;

the second is the electric field due to the motion of the magnet. The charge thus moves according to Newton's second law,

$$M_e \frac{d\mathbf{v}_e}{dt} = e\mathbf{E}_m + \frac{e}{c}\mathbf{v}_e \times \mathbf{B}_m,$$

where the right-hand side is the appropriate Lorentz force which the magnet exerts on the charge.

Now consider the magnet. Although Lagrange's equations can be obtained in the usual way, it is simpler to observe that the Lagrangian is unchanged on interchanging the subscripts e and m everywhere (but the quantity $\mathbf{A}$ changes sign). Applying this prescription to Lagrange's equations for the charge, we obtain Lagrange's equations for the magnet,

$$M_m \frac{d\mathbf{v}_m}{dt} = -\frac{1}{c}\frac{d\mathbf{m}}{dt} \times \left(e\frac{\mathbf{r}_m - \mathbf{r}_e}{|\mathbf{r}_m - \mathbf{r}_e|^3} \right) - \frac{e}{c}(\mathbf{v}_m - \mathbf{v}_e) \times (\nabla_m \times \mathbf{A}).$$

Note that

$$M_m \frac{d\mathbf{v}_m}{dt} = -M_e \frac{d\mathbf{v}_e}{dt},$$

so the total mechanical momentum $M_e\mathbf{v}_e + M_m\mathbf{v}_m$ is constant. The preceding equation needs to be rewritten so we can interpret it. First note that the quantity

$$\mathbf{E}_e = e\frac{\mathbf{r}_m - \mathbf{r}_e}{|\mathbf{r}_m - \mathbf{r}_e|^3}$$

in the first term on the right is the electric field due to the charge at the location of the magnet. Then in the second term set

$$e\mathbf{A} = -\mathbf{m} \times \mathbf{E}_e,$$

and hence find

$$e\nabla \times \mathbf{A} = -\nabla_m \times (\mathbf{m} \times \mathbf{E}_e) = (\mathbf{m} \cdot \nabla_m)\mathbf{E}_e.$$

Here we have used the vector identity $\nabla \times (\mathbf{a} \times \mathbf{b}) = \mathbf{a}(\nabla \cdot \mathbf{b}) - \mathbf{b}(\nabla \cdot \mathbf{a}) + (\mathbf{b} \cdot \nabla)\mathbf{a} - (\mathbf{a} \cdot \nabla)\mathbf{b}$ together with the fact that $\nabla_m \cdot \mathbf{E}_e = 0$. We then have

$$M_m \frac{d\mathbf{v}_m}{dt} = -\frac{1}{c}\frac{d\mathbf{m}}{dt} \times \mathbf{E}_e + (\mathbf{m} \cdot \nabla_m)\left(\frac{1}{c}\mathbf{v}_e \times \mathbf{E}_e - \frac{1}{c}\mathbf{v}_m \times \mathbf{E}_e\right).$$

The first term in the round brackets,

$$\mathbf{B}_e = \frac{1}{c}\mathbf{v}_e \times \mathbf{E}_e,$$

is the magnetic field due to the moving charge at the location of the magnet. We thus obtain

$$M_m \frac{d\mathbf{v}_m}{dt} = -\frac{1}{c}\frac{d\mathbf{m}}{dt} \times \mathbf{E}_e + (\mathbf{m}\cdot\nabla_m)\left(\mathbf{B}_e - \frac{1}{c}\mathbf{v}_m \times \mathbf{E}_e\right).$$

The term in round brackets is the magnetic field in the rest frame of the magnet. This is the correct equation of motion for the magnet, as is discussed more fully in *Lagrangian and Hamiltonian Mechanics*.

CHAPTER IV

THE PRINCIPLE OF STATIONARY ACTION OR HAMILTON'S PRINCIPLE

Exercise 4.01

Consider a modified brachistochrone problem in which the particle has non-zero initial speed v_0. Show that the brachistochrone is again a cycloid, but with cusp $h = v_0^2/2g$ higher than the initial point.

Solution

In this modified brachistochrone problem the speed of the particle when it has fallen a distance y is

$$v = \sqrt{v_0^2 + 2gy},$$

and the expression for the travel time becomes

$$\Delta t = \int_0^{x_1} \sqrt{\frac{1 + (dy/dx)^2}{v_0^2 + 2gy}}\, dx.$$

The integrand of this expression,

$$F(y, dy/dx) = \sqrt{\frac{1 + (dy/dx)^2}{v_0^2 + 2gy}},$$

does not depend explicitly on x, so for the required curve the quantity

$$H = \frac{\partial F}{\partial(dy/dx)} \frac{dy}{dx} - F = \frac{-1}{\sqrt{(v_0^2 + 2gy)(1 + (dy/dx)^2)}}$$

is constant. It is convenient to set $v_0^2 = 2gh$ and then set

$$(h + y)(1 + (dy/dx)^2) = 2a$$

where h and a are constants with the dimensions of "length." Rearranging this and integrating, we find

$$\int_0^y \sqrt{\frac{h+y}{2a-h-y}}\,dy = x.$$

The y-integration can be performed by setting

$$h + y = a(1-\cos\phi) \qquad \text{and} \qquad dy = a\sin\phi\, d\phi$$

to obtain

$$x = a\int_{\phi_0}^{\phi} (1-\cos\phi)\,d\phi = a\big[(\phi - \sin\phi) - (\phi_0 - \sin\phi_0)\big]$$

where ϕ_0 is determined by the condition $h = a(1-\cos\phi_0)$. The preceding two equations describe a cycloid which has a cusp at $(\phi = 0, x = -a(\phi_0 - \sin\phi_0), y = -h)$ and which passes through the initial point $(\phi = \phi_0, x = 0, y = 0)$ (Fig. 1). The constant a must be chosen so that the cycloid passes through the final point (x_1, y_1).

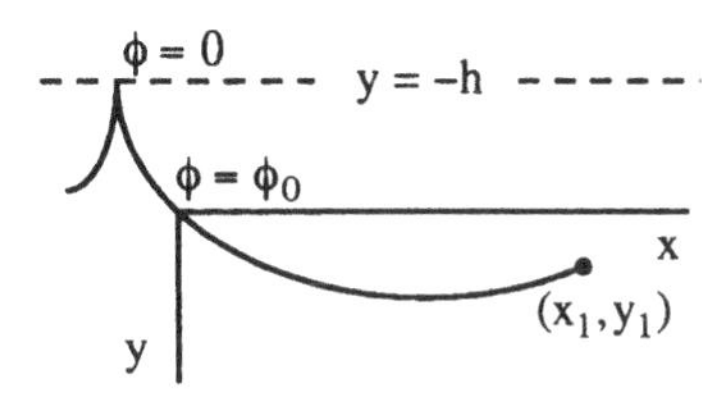

Ex. 4.01, Fig. 1

Exercise 4.02

A bead of mass m slides without friction along a wire bent in the shape of a cycloid

$$x = a(\phi - \sin\phi) \qquad y = a(1-\cos\phi).$$

Gravity g acts vertically down, parallel to the y axis.

(a) Find the displacement s along the cycloid, measured from the bottom, in terms of the parameter ϕ.

(b) Write down the Lagrangian using s as generalized coordinate, and show that the motion is simple harmonic in s with period independent of amplitude. Thus the time required for the bead, starting from rest, to slide from any point on the cycloid to the bottom is independent of the starting point. What is this time?

Solution

(a) The element of distance ds along the cycloid is given by

$$\begin{aligned}(ds)^2 &= (dx)^2 + (dy)^2 \\ &= a^2\left[(1-\cos\phi)^2 + \sin^2\phi\right](d\phi)^2 \\ &= 2a^2(1-\cos\phi)(d\phi)^2 \\ &= 4a^2\sin^2(\phi/2)(d\phi)^2 ,\end{aligned}$$

so

$$ds = 2a\sin(\phi/2)d\phi .$$

The displacement s, measured along the cycloid from the bottom ($\phi = \pi$), is thus

$$s = 2a\int_{\pi}^{\phi}\sin(\phi/2)\,d\phi = -4a\cos(\phi/2)$$

and ranges from $-4a$ to $4a$ as we move along the cycloid from cusp ($\phi = 0$) to cusp ($\phi = 2\pi$).

(b) The kinetic energy of a bead of mass m which slides along the cycloid is

$$T = \tfrac{1}{2}m\dot{s}^2 .$$

The gravitational potential energy of the bead, measured from the bottom of the cycloid, is

$$V = mg(2a - y) = mga(1+\cos\phi) = 2mga\cos^2(\phi/2) = \tfrac{1}{2}m(g/4a)s^2 .$$

The Lagrangian is thus

$$L = \tfrac{1}{2}m\dot{s}^2 - \tfrac{1}{2}m(g/4a)s^2 ,$$

and Lagrange's equation is

$$m\ddot{s} = -m(g/4a)s .$$

This is the equation of motion of a simple harmonic oscillator with angular frequency $\omega = \sqrt{g/4a}$ and period $\tau = 2\pi/\omega = 2\pi\sqrt{4a/g}$. The time required for the bead, starting from rest, to slide from any point on the cycloid to the bottom is one-quarter period, $\pi\sqrt{a/g}$, and is independent of the start point.

Exercise 4.03

Novelists have long been fascinated with the idea of a worldwide rapid transit system consisting of subterranean passages crisscrossing the earth.[1] Public interest in subterranean travel rose sharply when *Time* magazine[2] commented on a paper by Paul W. Cooper, "Through the Earth in Forty Minutes".[3] This paper, while repeating some earlier work,[4] served as a catalyst for a number of other papers on the subject[5] to which you may wish to refer in working the present exercise. Take the gravitational potential within the earth to be $\frac{1}{2}(g/R)r^2$ where g is the gravitational field at the surface and R is the radius of the earth (thereby neglecting the non-uniform density of the earth).

(a) First show that a particle starting from rest and sliding without friction through a *straight* tunnel connecting two points on the surface of the earth executes simple harmonic motion, and that the time to slide from one end to the other is $\tau_0 = \pi\sqrt{R/g}$ ($\approx 42.2\,\text{min}$) independent of the location of the end points.

(b) Now consider the curve $r(\theta)$ the tunnel must follow such that the time for the particle to slide from one end to the other is minimum. Set up the appropriate variational principle, and show that

$$\frac{r^2}{\sqrt{(dr/d\theta)^2 + r^2}\sqrt{R^2 - r^2}} = \frac{r_0}{\sqrt{R^2 - r_0^2}}$$

is a first integral of the resulting Euler-Lagrange equation. Here $r = r_0$ at the bottom of the tunnel (r_0 is the minimum distance to the center of the earth). Rearrange this and integrate to obtain the equation of the curve,

$$\theta = \tan^{-1}\left(\frac{R}{r_0}\sqrt{\frac{r^2 - r_0^2}{R^2 - r^2}}\right) - \frac{r_0}{R}\tan^{-1}\left(\sqrt{\frac{r^2 - r_0^2}{R^2 - r^2}}\right),$$

where θ is measured from the bottom of the tunnel. The angular separation between the end points on the surface of the earth is thus given by

$$\Delta\theta = \pi(1 - r_0/R).$$

(c) Introduce a parameter ϕ with

---

[1] See Martin Gardner, *Scientific American*, September 1965, pp. 10-12, commenting on an article by L. K. Edwards, "High-Speed Tube Transportation," *Scientific American*, August 1965, pp. 30-40.
[2] *Time*, February 11, 1966, pp. 42-43.
[3] Paul W. Cooper, "Through the Earth in Forty Minutes," Am. J. Phys. **34**, 68-70 (1966).
[4] See Philip G. Kirmser, "An Example of the Need for Adequate References," Am. J. Phys. **34**, 701 (1966).
[5] Giulio Venezian, "Terrestrial Brachistochrone," Am. J. Phys. **34**, 701 (1966); Russell L. Mallett, "Comments on 'Through the Earth in Forty Minutes'," Am. J. Phys. **34**, 702 (1966); L. Jackson Laslett, "Trajectory for Minimum Transit Time Through the Earth," Am. J. Phys. **34**, 702-703 (1966); Paul W. Cooper, "Further Commentary on 'Through the Earth in Forty Minutes'," Am. J. Phys. **34**, 703-704 (1966).

$$\tan\frac{\phi}{2} = \sqrt{\frac{r^2 - r_0^2}{R^2 - r^2}},$$

so $\phi = 0$ at the bottom and $\phi = \pm\pi$ at the ends of the tunnel. Show that the equation of the curve takes the form

$$r^2 = \tfrac{1}{2}\left(R^2 + r_0^2\right) - \tfrac{1}{2}\left(R^2 - r_0^2\right)\cos\phi$$

$$\theta = \tan^{-1}\left(\frac{R}{r_0}\tan\frac{\phi}{2}\right) - \frac{r_0}{2R}\phi .$$

Show that this is the equation of a hypocycloid, which is the curve traced by a point on the circumference of a circle which rolls without slipping on another circle.

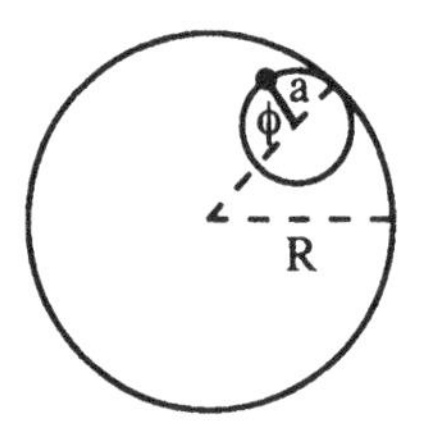

In this case the larger circle is the great circle route, of radius R, connecting the end points on the surface of the earth, and the smaller circle has radius $a = \frac{1}{2}(R - r_0)$ (its circumference is thus the distance between the end points on the surface). The parameter ϕ is the angle shown in the figure.

(d) Now consider the time dependence of the variables. Show in particular that ϕ varies linearly with time, $\phi = 2\pi(t/\tau)$, where $\tau = \tau_0\sqrt{1 - (r_0/R)^2}$ is the time to slide through the minimum-time-tunnel from one end to the other. Compare τ with τ_0 for end points 700 km apart on the surface.

Solution

(a) For a particle of mass m which starts from rest at the surface and which slides without friction through a tunnel through the earth, conservation of energy gives

$$\tfrac{1}{2}mv^2 + \tfrac{1}{2}m(g/R)r^2 = 0 + \tfrac{1}{2}m(g/R)R^2 .$$

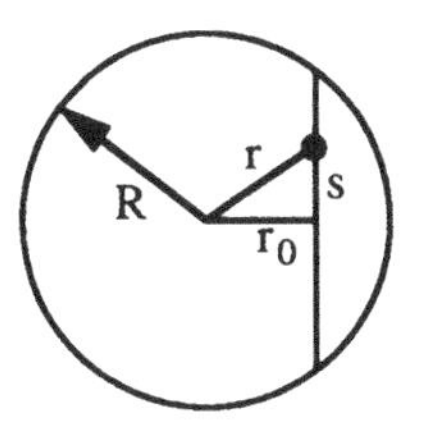

Ex. 4.03, Fig. 1

First suppose that the tunnel is straight and let s denote the displacement along the tunnel from the bottom, where r assumes its minimum value r_0 (Fig. 1). We then have

$$v = \dot{s} \quad \text{and} \quad r^2 = s^2 + r_0^2,$$

and the conservation of energy equation becomes

$$\tfrac{1}{2}m\dot{s}^2 + \tfrac{1}{2}m(g/R)s^2 = \tfrac{1}{2}m(g/R)(R^2 - r_0^2).$$

The left-hand side of this equation, as a function of s, has the same form as the expression for the total energy of a simple harmonic oscillator with angular frequency

$$\omega = \sqrt{g/R}\,.$$

The motion in s is thus simple harmonic. The time τ_0 for the particle to slide from one end of the tunnel to the other is half a period,

$$\tau_0 = \pi/\omega = \pi\sqrt{R/g} = 42.2 \text{ min}.$$

(b) Now consider the curve the tunnel must follow to make the travel time

$$\Delta t = \int_{t_0}^{t_1} \frac{ds}{v}$$

for the particle a minimum. It is convenient to use plane polar coordinates (r,θ), in terms of which the element of distance ds is

$$(ds)^2 = (dr)^2 + r^2(d\theta)^2.$$

The speed v of the particle is given by energy conservation (see part (a)),

$$v^2 = (g/R)(R^2 - r^2).$$

The expression for the travel time becomes

$$\Delta t = \sqrt{\frac{R}{g}} \int_{\theta_0}^{\theta_1} \sqrt{\frac{(dr/d\theta)^2 + r^2}{R^2 - r^2}}\, d\theta.$$

We wish to find the curve $r = r(\theta)$ which makes this a minimum. The integrand,

$$F = \sqrt{\frac{R}{g}} \sqrt{\frac{(dr/d\theta)^2 + r^2}{R^2 - r^2}},$$

does not contain the independent variable θ explicitly, so for the required curve the quantity

$$H = \frac{\partial F}{\partial (dr/d\theta)} \frac{dr}{d\theta} - F = -\sqrt{\frac{R}{g}} \frac{1}{\sqrt{R^2 - r^2}} \frac{r^2}{\sqrt{(dr/d\theta)^2 + r^2}}$$

is constant. We set

$$\frac{1}{\sqrt{R^2 - r^2}} \frac{r^2}{\sqrt{(dr/d\theta)^2 + r^2}} = \frac{r_0}{\sqrt{R^2 - r_0^2}}.$$

At the bottom of the tunnel $dr/d\theta = 0$ and this equation gives $r = r_0$, so the constant r_0 is the minimum value of r. Solving for $dr/d\theta$, we find

$$\frac{dr}{d\theta} = R \frac{r}{r_0} \sqrt{\frac{r^2 - r_0^2}{R^2 - r^2}}.$$

Rearranging this and integrating, we then find

$$R\theta = \int_{r_0}^{r} \frac{r_0}{r} \sqrt{\frac{R^2 - r^2}{r^2 - r_0^2}}\, dr$$

where we have chosen the constant of integration so that $\theta = 0$ at the bottom of the tunnel (Fig. 2).

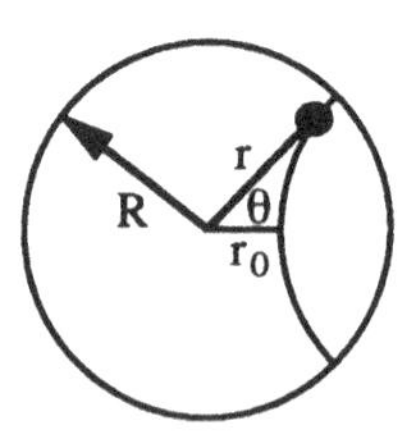

Ex. 4.03, Fig. 2

To do the integration, we set

$$u = \sqrt{\frac{r^2 - r_0^2}{R^2 - r^2}}.$$

The variable u is zero for $r = r_0$ and it tends to $\pm\infty$ for $r \to R$. The integral becomes

$$R\theta = r_0 \int_0^u \left(\frac{1}{(r_0/R)^2 + u^2} - \frac{1}{1 + u^2} \right) du = R \tan^{-1}\left(\frac{R}{r_0} u \right) - r_0 \tan^{-1} u,$$

so the equation of the curve the minimum time tunnel must follow is

$$\theta = \tan^{-1} \frac{R}{r_0} \sqrt{\frac{r^2 - r_0^2}{R^2 - r^2}} - \frac{r_0}{R} \tan^{-1} \sqrt{\frac{r^2 - r_0^2}{R^2 - r^2}}.$$

The ends of the tunnel are at $r = R$, so the angular separation of the ends is

$$\Delta\theta = \pi(1 - r_0/R).$$

(c) It is convenient to set

$$\tan\frac{\phi}{2} = u = \sqrt{\frac{r^2 - r_0^2}{R^2 - r^2}},$$

where the parameter ϕ is zero at the bottom and $\pm\pi$ at the ends of the tunnel. Solving for r, we find

$$r^2 = R^2 \sin^2(\phi/2) + r_0^2 \cos^2(\phi/2) = \tfrac{1}{2}(R^2 + r_0^2) - \tfrac{1}{2}(R^2 - r_0^2)\cos\phi.$$

Further, from the equation of the tunnel in part (b) we have

$$\theta = \tan^{-1}\left(\frac{R}{r_0} \tan\frac{\phi}{2} \right) - \frac{r_0}{2R}\phi.$$

These two equations, which give r and θ as functions of the parameter ϕ, are another way of describing the curve which the minimum time tunnel must follow. We wish to show that this curve is a hypocycloid, which is the curve traced by a point on the rim of a circle which rolls on another circle. In this case a smaller circle of radius $a = \tfrac{1}{2}(R - r_0)$ rolls inside a larger circle of radius R (Fig. 3).

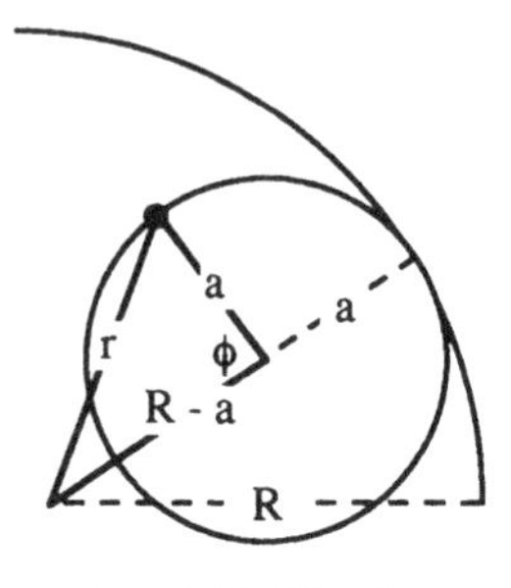

Ex. 4.03, Fig. 3

If we apply the trigonometric cosine law to the triangle in Fig. 3, we have

$$r^2 = (R - a)^2 + a^2 - 2a(R - a)\cos\phi = \tfrac{1}{2}(R^2 + r_0^2) - \tfrac{1}{2}(R^2 + r_0^2)\cos\phi,$$

which is one of the preceding equations. To obtain the second equation, the relation between θ and ϕ, we proceed as follows. Let θ_0 denote the angle to the center of the smaller rolling circle and let θ denote the angle to the point on the circumference of the smaller circle which traces the curve (Fig. 4)

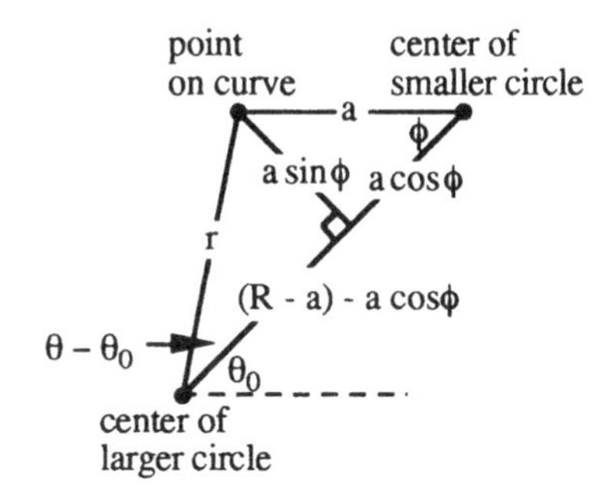

Ex. 4.03, Fig. 4

Then from Fig. 4

$$\tan(\theta - \theta_0) = \frac{a\sin\phi}{(R-a) - a\cos\phi} = \frac{(R - r_0)\sin\phi}{R(1-\cos\phi) + r_0(1+\cos\phi)} = \frac{(R - r_0)\tan(\phi/2)}{R\tan^2(\phi/2) + r_0}.$$

Now note that

$$\tan(\theta - \theta_0 + (\phi/2)) = \frac{\tan(\theta - \theta_0) + \tan(\phi/2)}{1 - \tan(\theta - \theta_0)\tan(\phi/2)} = (R/r_0)\tan(\phi/2),$$

where to obtain the second equality we have substituted $\tan(\theta - \theta_0)$ from the preceding equation. The angle θ_0 is related to ϕ by the condition that the smaller circle *rolls* without slipping on the larger. This means that the arc $R\theta_0$ of the larger circle over which the point of contact moves must equal the arc $a\phi$ of the smaller circle,

$$R\theta_0 = a\phi = (R - r_0)(\phi/2),$$

and thus

$$-\theta_0 + (\phi/2) = (r_0/R)(\phi/2).$$

Substituting this into the preceding equation and rearranging, we obtain

$$\theta = \tan^{-1}\left(\frac{R}{r_0}\tan\frac{\phi}{2}\right) - \frac{r_0}{2R}\phi$$

which is the required relation between θ and ϕ.

(d) The time, measured from the bottom of the tunnel, is

$$t = \sqrt{\frac{R}{g}}\int_0^\theta \sqrt{\frac{(dr/d\theta)^2 + r^2}{R^2 - r^2}}\, d\theta.$$

We have seen in part (b) that

$$\frac{dr}{d\theta} = R\frac{r}{r_0}\sqrt{\frac{r^2 - r_0^2}{R^2 - r^2}}$$

along the minimum time tunnel, so the expression for the time becomes

$$t = \sqrt{\frac{R^2 - r_0^2}{gR}}\int_{r_0}^{r} \frac{r\,dr}{\sqrt{(R^2 - r^2)(r^2 - r_0^2)}}.$$

Now introduce again the parameter ϕ with

$$\tan\frac{\phi}{2} = \sqrt{\frac{r^2 - r_0^2}{R^2 - r^2}} \quad \text{and} \quad d\phi = \frac{2r\,dr}{\sqrt{(R^2 - r^2)(r^2 - r_0^2)}}.$$

Integration gives

$$t = \frac{\phi}{2}\sqrt{\frac{R^2 - r_0^2}{gR}},$$

so ϕ is a linear function of the time. The total time for the particle to travel from one end of the tunnel to the other is

$$\tau = \pi\sqrt{\frac{R^2 - r_0^2}{gR}} = \tau_0\sqrt{1-(r_0/R)^2},$$

so the parameter ϕ can be written

$$\phi = 2\pi(t/\tau).$$

Note that $\tau \approx 0$ for $r_0 \approx R$ (short tunnel near the surface of the earth) and that $\tau = \tau_0$ for $r_0 = 0$ (straight tunnel through the center of the earth). For end points 700 km apart on the surface the angular separation is

$$\Delta\theta = 2\pi\frac{700}{40000} \quad \text{which gives} \quad 1-\frac{r_0}{R} = \frac{2\times 700}{40000}.$$

Hence the travel time along the minimum time tunnel connecting these two points is

$$\tau = \tau_0\sqrt{1-\left(\frac{r_0}{R}\right)^2} \approx \tau_0\sqrt{\frac{4\times 700}{40000}} = 0.26\tau_0 = 11 \text{ min}.$$

Exercise 4.04

An instructive exercise in the calculus of variations is the "minimum surface of revolution problem":

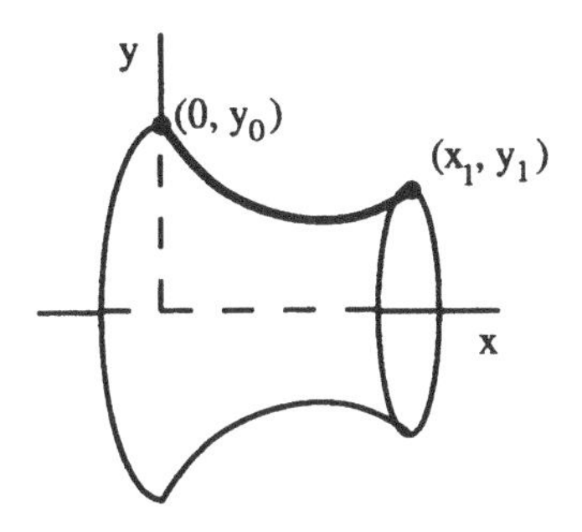

(a) Find the plane curve $y = y(x)$ joining two points $(0,y_0)$ and (x_1,y_1) such that the area of the surface formed by rotating the curve about the x-axis is minimum (the Euler-Lagrange answer is $y = a\cosh((x-b)/a)$, where a and b must be chosen so that the curve passes through the end points).
(b) Using a computer or otherwise, draw representative members of the (one-parameter) family of such curves which start at $(0,1)$. Hence convince yourself that if the final point is near the y-axis, *two* Euler-Lagrange curves pass through the given end points, whereas if the final point is near the x-axis, *no* Euler-Lagrange curves pass through the given end points.
(c) In this latter case the solution is the discontinuous Goldschmidt solution composed of straight line segments $(0,1) \to (0,0) \to (x_1,0) \to (x_1,y_1)$. In the region where there are two Euler-Lagrange solutions, calculate and compare the area given by the Goldschmidt solution with the areas given by the Euler-Lagrange solutions. Which of the three gives minimum area? Does this depend on where in the region the end point lies? (This last part is difficult; for guidance see Gilbert Ames Bliss, *Calculus of Variations*, (published for The Mathematical Association of America by The Open Court Publishing Company, Chicago, Illinois, 1925), Chap. IV, pp. 85-127.)

Solution

(a) Consider the surface of revolution generated by rotating the curve $y = y(x)$ around the x-axis. A little strip around the surface has an area

$$dA = 2\pi y\, ds = 2\pi y\sqrt{1+(dy/dx)^2}\, dx,$$

so the total surface area between limits x_0 and x_1 is

$$A = 2\pi\int_{x_0}^{x_1} y\sqrt{1+(dy/dx)^2}\, dx.$$

We wish to find the curve $y = y(x)$ which makes this surface area a minimum. The integrand

$$F = 2\pi y\sqrt{1 + (dy/dx)^2}$$

does not depend explicitly on the independent variable x, so for the required curve the quantity

$$H = \frac{\partial F}{\partial(dy/dx)}\frac{dy}{dx} - F = -\frac{2\pi y}{\sqrt{1 + (dy/dx)^2}}$$

is constant. We set

$$\frac{y}{\sqrt{1 + (dy/dx)^2}} = a.$$

The constant a is the minimum value of y (where $dy/dx = 0$). Solving for dy/dx, we find

$$\frac{dy}{dx} = \frac{1}{a}\sqrt{y^2 - a^2}.$$

Rearranging this and integrating, we then obtain

$$\frac{x - b}{a} = \int_a^y \frac{dy}{\sqrt{y^2 - a^2}} = \cosh^{-1}\frac{y}{a}.$$

The constant of integration b is the value of x at which y is a minimum. Thus, according to Euler-Lagrange, the curve which generates the surface of revolution with minimum area is

$$y = a\cosh(x - b)/a.$$

Such a curve is called a "catenary" (Fig. 1). The constants a and b must be chosen so that the catenary passes through the specified end points. As we shall see in part (b), this is not always possible. What has gone wrong? The difficulty is that the Euler-Lagrange approach implicitly assumes that the required curve is smooth, and this is not always the case. It turns out that for the region where there is no Euler-Lagrange curve, and indeed even in part of the region where there are Euler-Lagrange curves, the solution is the non-smooth Goldschmidt curve.

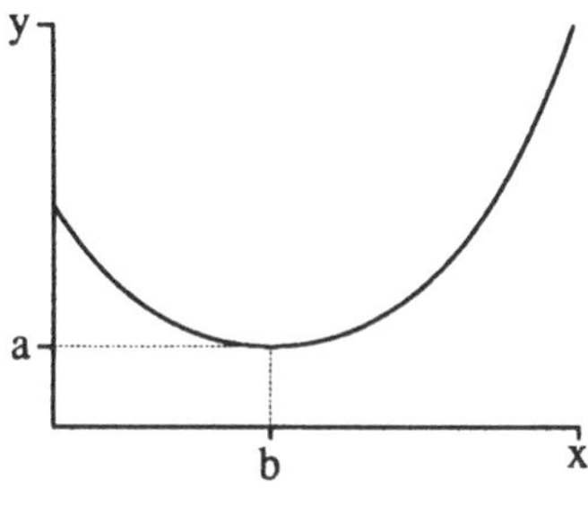

Ex. 4.04, Fig. 1

(b) For catenaries which start at (0,1) the constants a and b are related by

$$1 = a\cosh(b/a).$$

Since "cosh" is always greater than or equal to 1, the constant a lies between 0 and 1, and

$$b = \pm a\cosh^{-1}(1/a).$$

The family of catenaries obtained as a is varied from 0 to 1 is shown in Fig. 2. As we can see, this family has an envelope such that two catenaries pass through any point above the envelope, and no catenaries pass through any point below the envelope.

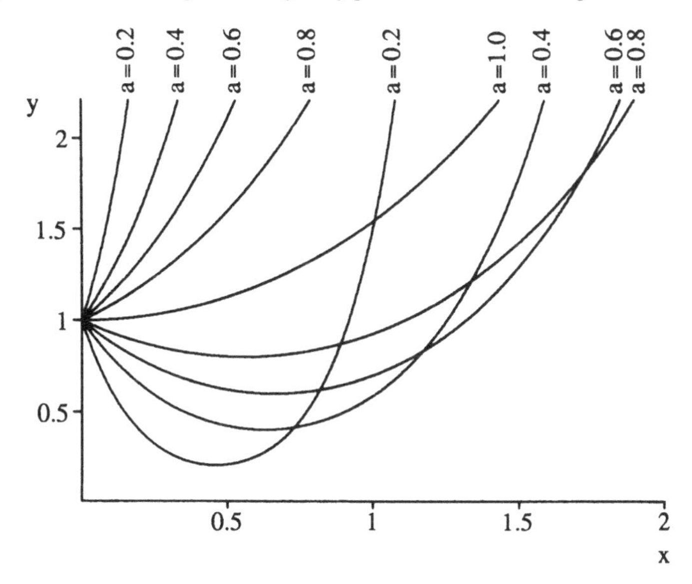

Ex. 4.04, Fig. 2

(c) The area of the surface of revolution generated by rotating one of these catenaries around the x-axis is

$$A = 2\pi\int_0^{x_1} y\sqrt{1+(dy/dx)^2}\,dx$$
$$= 2\pi a\int_0^{x_1} \cosh^2((x-b)/a)\,dx$$
$$= \pi a^2\left[\frac{x_1}{a} + \frac{1}{2}\sinh 2\frac{x_1-b}{a} + \frac{1}{2}\sinh 2\frac{b}{a}\right]$$

where $b = \pm a\cosh^{-1}(1/a)$ still. We have seen that two such catenaries pass through any end point above the envelope. How do the areas of the surfaces of revolution generated by these compare? Also, how do they compare with the area

$$A_{\text{Gold}} = \pi(1)^2 + \pi y_1^2$$

generated by the non-smooth Goldschmidt curve?

We shall not attempt to give a complete solution to this problem. Rather, we take a few end points and see what happens, the aim being to bring out at least some of the general features of the problem. In particular, we take as end points the intersections of the $a = 0.2$ catenary (the "lower" catenary) with the $a = 0.4, 0.6, 0.8,$ and 1.0 catenaries (the "upper" catenary). The end points and the areas of the surfaces of revolution generated by the upper and lower catenaries , and by the Goldschmidt curve, are given in Table I.

Table I: Areas of surfaces of revolution

$a = 0.2$ with	$x_1 =$	$y_1 =$	$A_{\text{upper}} =$	$A_{\text{lower}} =$	$A_{\text{Gold}} =$
$a = 0.4$	0.729359	0.413241	3.930575	4.005856	**3.678077**
$a = 0.6$	0.818405	0.621255	4.370380	4.740310	**4.354114**
$a = 0.8$	0.888642	0.870795	**5.059222**	5.955004	5.523810
$a = 1.0$	1.005808	1.549932	**8.925927**	11.194006	10.688607

We see that: (a) the area generated by the upper catenary is less than that generated by the lower catenary in all cases, and (b) the area generated by the Goldschmidt curve is less than that generated by the upper catenary if the end point is "near" the envelope but is greater than the area generated by the upper catenary if the end point is "far from" the envelope. These turn out to be general features of the minimum surface of revolution problem.

Exercise 4.05

The motion of a "free" particle of mass m on a surface is described by Lagrange's equations with Lagrangian $L = T = \frac{1}{2}m(ds/dt)^2$. Show that the resulting equations of motion are the equations for a geodesic, along which the particle moves at constant speed ds/dt.

Solution

The Lagrangian is

$$L = \frac{1}{2}m\left(\frac{ds}{dt}\right)^2 = \frac{1}{2}mg_{\rho\sigma}\frac{dx^\rho}{dt}\frac{dx^\sigma}{dt}$$

where the x^ρ ($\rho = 1,2$) are two independent coordinates on the surface and the $g_{\rho\sigma}(x)$ are the components of the metric tensor. We have

$$\frac{\partial L}{\partial(dx^\alpha/dt)} = mg_{\alpha\rho}\frac{dx^\rho}{dt}, \qquad \frac{\partial L}{\partial x^\alpha} = \frac{1}{2}m\frac{\partial g_{\rho\sigma}}{\partial x^\alpha}\frac{dx^\rho}{dt}\frac{dx^\sigma}{dt},$$

$$\frac{d}{dt}\left(\frac{\partial L}{\partial(dx^\alpha/dt)}\right) = mg_{\alpha\rho}\frac{d^2x^\rho}{dt^2} + m\frac{\partial g_{\alpha\rho}}{\partial x^\sigma}\frac{dx^\rho}{dt}\frac{dx^\sigma}{dt},$$

so Lagrange's equations are

$$g_{\alpha\rho}\frac{d^2x^\rho}{dt^2} = -\frac{1}{2}\left(2\frac{\partial g_{\alpha\rho}}{\partial x^\sigma} - \frac{\partial g_{\rho\sigma}}{\partial x^\alpha}\right)\frac{dx^\rho}{dt}\frac{dx^\sigma}{dt} = -\Gamma_{\alpha,\rho\sigma}\frac{dx^\rho}{dt}\frac{dx^\sigma}{dt}$$

where the $\Gamma_{\alpha,\rho\sigma}(x)$ are the components of the Christoffel symbol. These equations are the same as the equations of a geodesic found in *Lagrangian and Hamiltonian Mechanics*, page 66. Further, since the Lagrangian does not depend explicitly on t, the quantity

$$H = \frac{\partial L}{\partial(dx^\alpha/dt)}\frac{dx^\alpha}{dt} - L = \frac{1}{2}mg_{\rho\sigma}\frac{dx^\rho}{dt}\frac{dx^\sigma}{dt} = \frac{1}{2}m\left(\frac{ds}{dt}\right)^2$$

is constant in time; that is, the speed ds/dt of the particle is constant in time.

Exercise 4.06

Find and solve the equations for geodesics on a plane, using plane polar coordinates (r,ϕ) in terms of which the element ds of distance is given by $ds^2 = dr^2 + r^2 d\phi^2$.

Solution

The non-zero components of the metric tensor are

$$g_{rr} = 1, \qquad g_{\phi\phi} = r^2.$$

The non-zero components of the Christoffel symbol are then

$$\Gamma_{\phi,\phi r} = \Gamma_{\phi,r\phi} = -\Gamma_{r,\phi\phi} = \frac{1}{2}\frac{\partial g_{\phi\phi}}{\partial r} = r,$$

and the equations for a geodesic become

$$\frac{d^2r}{ds^2} = r\left(\frac{d\phi}{ds}\right)^2, \qquad r^2\frac{d^2\phi}{ds^2} = -2r\frac{dr}{ds}\frac{d\phi}{ds}.$$

The second of these equations can be written

$$\frac{d}{ds}\left(r^2\frac{d\phi}{ds}\right) = 0$$

and gives

$$r^2\frac{d\phi}{ds} = r_0$$

where r_0 is a constant. Using this to eliminate $d\phi/ds$ from the first equation, we find

$$\frac{d^2r}{ds^2} = \frac{r_0^2}{r^3} = -\frac{d}{ds}\left(\frac{r_0^2}{2r^2}\right).$$

This is like the equation of motion of a particle of unit mass in a potential $r_0^2/2r^2$, so "energy conservation" yields

$$\frac{1}{2}\left(\frac{dr}{ds}\right)^2 + \frac{r_0^2}{2r^2} = \text{constant}.$$

The value of the constant can be found from the constraint

$$1 = \left(\frac{dr}{ds}\right)^2 + r^2\left(\frac{d\phi}{ds}\right)^2 = \left(\frac{dr}{ds}\right)^2 + \frac{r_0^2}{r^2};$$

it thus equals 1/2. Rearranging and integrating, we obtain

$$s = \int_{r_0}^{r} \frac{r\,dr}{\sqrt{r^2 - r_0^2}} = \sqrt{r^2 - r_0^2}$$

where we have chosen the constant of integration so that $s = 0$ at $r = r_0$. This can be written

$$r^2 = r_0^2 + s^2.$$

Substituting the result into the ϕ-equation, we find

$$\frac{d\phi}{ds} = \frac{r_0}{r^2} = \frac{r_0}{r_0^2 + s^2}.$$

Integrating, we obtain

$$\phi = \int_0^s \frac{r_0\,ds}{r_0^2 + s^2} = \tan^{-1}\left(\frac{s}{r_0}\right)$$

where we have chosen $\phi = 0$ at $s = 0$. This can be written

$$s = r_0 \tan\phi.$$

The two equations $r^2 = r_0^2 + s^2$ and $s = r_0 \tan\phi$, which express the polar coordinates (r, ϕ) in terms of a parameter s, are the equations of a straight line (Fig. 1). If desired, the parameter s can be eliminated to give the explicit relation $r\cos\phi = r_0$ between the coordinates.

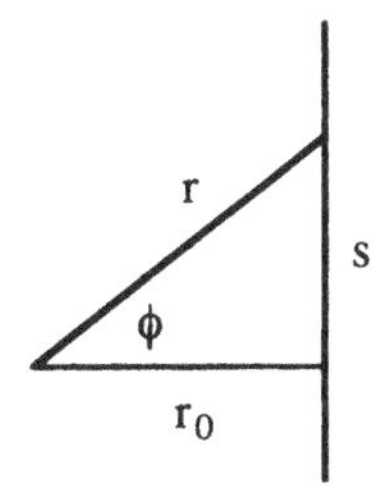

Ex. 4.06, Fig. 1

Exercise 4.07

(a) Find and solve the equations for geodesics on the surface of a cone of half angle α, using as coordinates the distance r from the apex of the cone and the azimuthal angle ϕ.
(b) Show that if the cone is cut along a line ϕ = constant and flattened out onto a plane, the geodesics become straight lines.

Solution

(a) The distance ds between two neighboring points on the surface of the cone is given by

$$(ds)^2 = (dr)^2 + r^2 \sin^2\alpha\,(d\phi)^2.$$

This has the same form as the expression for $(ds)^2$ in Exercise 4.06, except that the ϕ in Exercise 4.06 is here replaced by $\phi\sin\alpha$. As a result, the calculations and indeed the equation for a geodesic, $r\cos\phi = r_0$, in Exercise 4.06 can be easily modified to yield the equation for a geodesic,

$$r\cos(\phi\sin\alpha) = r_0,$$

in the present problem. We have taken $\phi = 0$ at minimum radius $r = r_0$; this geodesic can, of course, be rotated about the axis of the cone to give other geodesics.

(b) Now cut the cone along a line $\phi = \phi_0$ and flatten it out onto a plane. Straight lines ϕ = constant on the cone become radial straight lines on the plane. Circles r = constant on the cone become circular arcs on the plane; "arcs" because the distance around these circles on the cone is $2\pi r\sin\alpha$, so the angle they subtend on the plane is $2\pi\sin\alpha$. The two sides of the cut become a wedge of angle $2\pi(1-\sin\alpha)$ which is cut out of the plane. If we introduce polar coordinates (R, Φ) on the plane, then

$$R = r \quad \text{and} \quad \Phi = \phi\sin\alpha.$$

We have taken $\Phi = 0$ at $\phi = 0$. As ϕ ranges from $-(2\pi - \phi_0)$ through 0 to ϕ_0, the angle Φ ranges from $-(2\pi - \phi_0)\sin\alpha$ through 0 to $\phi_0\sin\alpha$ (Fig. 1).

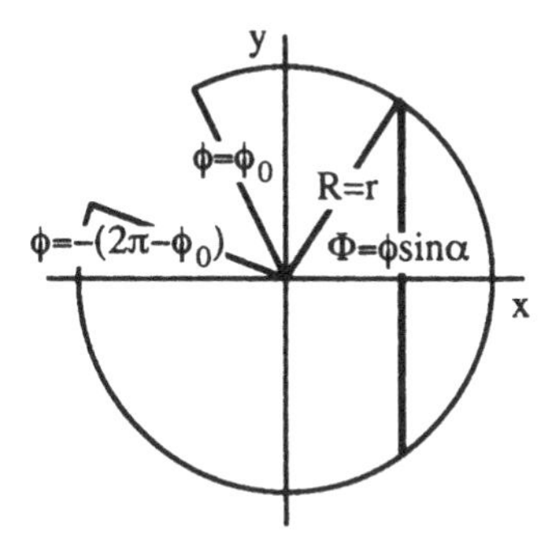

Ex. 4.07, Fig. 1

The equation for a geodesic on a cone, when the cone is cut and flattened out onto a plane, becomes

$$R\cos\Phi = r_0.$$

As we have seen in Exercise 4.06, this is the equation of a straight line on the plane.

Exercise 4.08

Consider two points on the surface of a sphere. Without loss we may take them on the equator at $(\theta = \pi/2, \phi = 0)$ and $(\theta = \pi/2, \phi = \alpha)$. The geodesics joining these points are the two arcs of the equator. Nearby curves can be represented by

$$\theta = \pi/2 + \sum_{n=1}^{\infty} a_n \sin(\pi n\phi/\alpha)$$

where the deviation from the equator has been represented by a Fourier series chosen to vanish at the end points. Evaluate the distance between the two points along such a curve, valid to second order in the small quantities a. Show that for $0 < \alpha < \pi$ the distance is always longer than the distance along the equator, whereas for $\pi < \alpha < 2\pi$ there are nearby curves for which the distance is shorter than that along the equator, as well as ones for which the distance is longer than that along the equator.

Solution

The distance ds between two neighboring points on the surface of a sphere of radius R is given by

$$(ds)^2 = R^2\left[(d\theta)^2 + \sin^2\theta(d\phi)^2\right]$$

where θ and ϕ are the spherical polar coordinates "co-latitude" and "longitude." We wish to find the distance along the curve

$$\theta = \frac{\pi}{2} + \sum_{n=1}^{\infty} a_n \sin\frac{n\pi\phi}{\alpha}$$

between the points ($\phi = 0, \theta = \pi/2$) and ($\phi = \alpha, \theta = \pi/2$). For this we need

$$d\theta = \sum_{n=1}^{\infty} a_n \frac{n\pi}{\alpha} \cos\frac{n\pi\phi}{\alpha} d\phi,$$

$$(d\theta)^2 = \sum_{n=1}^{\infty}\sum_{m=1}^{\infty} a_n a_m \frac{n\pi}{\alpha}\frac{m\pi}{\alpha} \cos\frac{n\pi\phi}{\alpha} \cos\frac{m\pi\phi}{\alpha} (d\phi)^2,$$

and

$$\sin\theta = \cos\left(\sum_{n=1}^{\infty} a_n \sin\frac{n\pi\phi}{\alpha}\right) = 1 - \frac{1}{2}\sum_{n=1}^{\infty}\sum_{m=1}^{\infty} a_n a_m \sin\frac{n\pi\phi}{\alpha} \sin\frac{m\pi\phi}{\alpha} + \cdots,$$

$$\sin^2\theta = 1 - \sum_{n=1}^{\infty}\sum_{m=1}^{\infty} a_n a_m \sin\frac{n\pi\phi}{\alpha} \sin\frac{m\pi\phi}{\alpha} + \cdots.$$

The element of distance is thus given by

$$(ds)^2 = R^2 (d\phi)^2 \left[1 + \sum_{n=1}^{\infty}\sum_{m=1}^{\infty} a_n a_m \left(\frac{n\pi}{\alpha}\frac{m\pi}{\alpha} \cos\frac{n\pi\phi}{\alpha} \cos\frac{m\pi\phi}{\alpha} - \sin\frac{n\pi\phi}{\alpha} \sin\frac{m\pi\phi}{\alpha}\right) + \cdots\right],$$

so

$$ds = R\, d\phi \left[1 + \frac{1}{2}\sum_{n=1}^{\infty}\sum_{m=1}^{\infty} a_n a_m \left(\frac{n\pi}{\alpha}\frac{m\pi}{\alpha} \cos\frac{n\pi\phi}{\alpha} \cos\frac{m\pi\phi}{\alpha} - \sin\frac{n\pi\phi}{\alpha} \sin\frac{m\pi\phi}{\alpha}\right) + \cdots\right].$$

To find the total distance along the curve, we integrate from $\phi = 0$ to $\phi = \alpha$. The integrals of $\cos\times\cos$ and $\sin\times\sin$ give zero for $m \neq n$ and $\alpha/2$ for $m = n$. We find

$$s = R\alpha \left[1 + \frac{1}{4}\sum_{n=1}^{\infty} a_n^2 \left(\frac{n^2\pi^2}{\alpha^2} - 1\right) + \cdots\right].$$

Since $0 \le \alpha < 2\pi$, the factor $n^2\pi^2/\alpha^2 - 1$ is positive for $n \ge 2$. We are interested here in curves of minimum distance and hence set $a_n = 0$ for $n \ge 2$. We then have

$$s = R\alpha\left[1 + \frac{1}{4}a_1^2\left(\frac{\pi^2}{\alpha^2} - 1\right) + \cdots\right].$$

If $0 \le \alpha < \pi$, the factor $\pi^2/\alpha^2 - 1$ is positive and all nearby curves are longer than the arc $R\alpha$ of the equator. If, however, $\pi < \alpha < 2\pi$, the factor $\pi^2/\alpha^2 - 1$ is negative and nearby curves of the form

$$\theta = \frac{\pi}{2} + a_1 \sin\frac{\pi\phi}{\alpha}$$

are shorter than the arc of the equator.

Exercise 4.09

(a) Evaluate the action $S[\mathbf{x}(t)]$ for a free particle along the path:
"from $(\mathbf{x}_0, t_0)$ to $(\mathbf{x}', t')$ at constant velocity, and then
from $(\mathbf{x}', t')$ to $(\mathbf{x}_1, t_1)$ at (a usually different) constant velocity."
(b) Consider S as a function of a parameter $\mathbf{x}'$. Show that minimum action results when $\mathbf{x}'$ is chosen so that the velocity from $(\mathbf{x}_0, t_0)$ to $(\mathbf{x}', t')$ is the same as that from $(\mathbf{x}', t')$ to $(\mathbf{x}_1, t_1)$, so that the full motion is at constant velocity.

Solution

The action $S[\mathbf{x}(t)] = \int_{t_0}^{t_1} L\,dt$ for a free particle, with Lagrangian $L = \frac{1}{2}m|\dot{\mathbf{x}}|^2$, along the given path (Fig. 1) is

$$S = \frac{m}{2}\frac{|\mathbf{x}' - \mathbf{x}_0|^2}{t' - t_0} + \frac{m}{2}\frac{|\mathbf{x}_1 - \mathbf{x}'|^2}{t_1 - t'}.$$

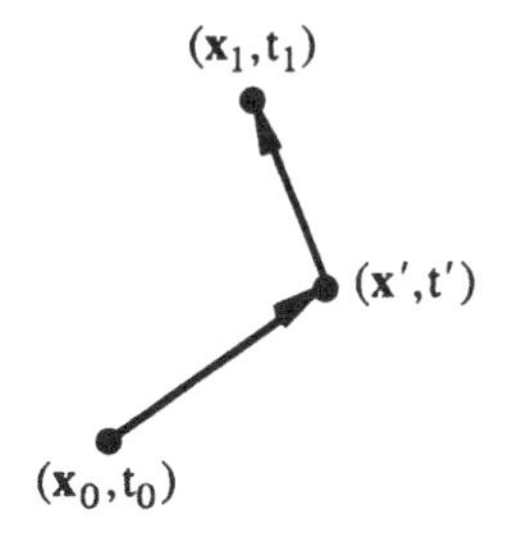

Ex. 4.09, Fig. 1

We adjust $\mathbf{x}'$ to give minimum S. This requires

$$\frac{\partial S}{\partial \mathbf{x}'} = m\frac{\mathbf{x}' - \mathbf{x}_0}{t' - t_0} - m\frac{\mathbf{x}_1 - \mathbf{x}'}{t_1 - t'} = 0,$$

so the velocity $\frac{\mathbf{x}' - \mathbf{x}_0}{t' - t_0}$ from $(\mathbf{x}_0, t_0)$ to $(\mathbf{x}', t')$ must equal the velocity $\frac{\mathbf{x}_1 - \mathbf{x}'}{t_1 - t'}$ from $(\mathbf{x}', t')$ to $(\mathbf{x}_1, t_1)$; the full motion is at constant velocity. This argument is easily generalized to allow for many intermediate points.

Exercise 4.10

Fermat's principle states that light travels from one point to another along the trajectory which makes the travel time a minimum.
(a) Use Fermat's principle to derive the law for the reflection of light from a mirror, namely

" angle of incidence = angle of reflection "

(b) Use Fermat's principle to derive Snell's law for the refraction of light passing from a medium in which the speed of light is c/n_0 to a medium in which the speed of light is c/n_1 (c is the speed of light in free space and n is the index of refraction), namely

$$n_0 \sin\phi_0 = n_1 \sin\phi_1.$$

Here ϕ_0 and ϕ_1 are the angles to the normal of the incident and refracted rays.

Solution

(a) Consider a light ray which travels from (x_0, y_0) to (x_1, y_1) via a point $(x,0)$ on the surface of a mirror (Fig. 1).

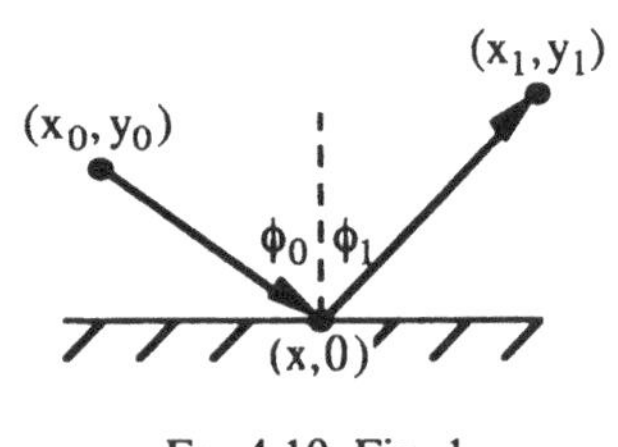

Ex. 4.10, Fig. 1

The travel time is

$$t = \frac{1}{c}\sqrt{(x - x_0)^2 + y_0^2} + \frac{1}{c}\sqrt{(x_1 - x)^2 + y_1^2}.$$

We adjust x to minimize this time. This requires

$$c\frac{\partial t}{\partial x} = \frac{x - x_0}{\sqrt{(x - x_0)^2 + y_0^2}} - \frac{x_1 - x}{\sqrt{(x_1 - x)^2 + y_1^2}} = 0.$$

Now

$$\frac{x - x_0}{\sqrt{(x - x_0)^2 + y_0^2}} = \sin\phi_0 \quad \text{and} \quad \frac{x_1 - x}{\sqrt{(x_1 - x)^2 + y_1^2}} = \sin\phi_1$$

where ϕ_0 and ϕ_1 are the angles, to the normal, of the incident and reflected rays. The preceding equation becomes

$$\sin\phi_0 = \sin\phi_1,$$

so

$$\phi_0 = \phi_1$$

which is the law of reflection from a mirror.

(b) Consider a light ray which travels from a point (x_0, y_0) in a medium in which the index of refraction is n_0 to a point (x_1, y_1) in a medium in which the index of refraction is n_1, via a point $(x, 0)$ on the interface (Fig. 2).

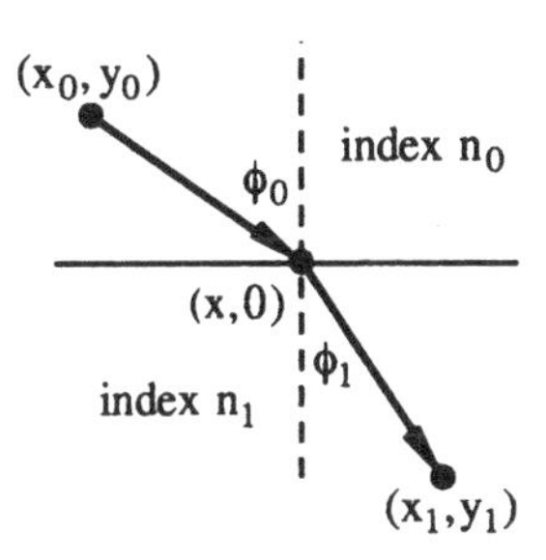

Ex. 4.10, Fig. 2

The travel time is

$$t = \frac{n_0}{c}\sqrt{(x - x_0)^2 + y_0^2} + \frac{n_1}{c}\sqrt{(x_1 - x)^2 + y_1^2}.$$

We adjust x to minimize this time. This requires

$$c\frac{\partial t}{\partial x} = n_0\frac{x - x_0}{\sqrt{(x - x_0)^2 + y_0^2}} - n_1\frac{x_1 - x}{\sqrt{(x_1 - x)^2 + y_1^2}} = 0.$$

Now

$$\frac{x - x_0}{\sqrt{(x - x_0)^2 + y_0^2}} = \sin\phi_0 \quad \text{and} \quad \frac{x_1 - x}{\sqrt{(x_1 - x)^2 + y_1^2}} = \sin\phi_1$$

where ϕ_0 and ϕ_1 are the angles, to the normal, of the incident and refracted rays. The preceding equation becomes

$$n_0 \sin\phi_0 = n_1 \sin\phi_1$$

which is Snell's law of refraction.

Exercise 4.11*

Jacobi's principle states that a particle of mass m and energy E in a potential $V(x,y,z)$ travels from one point to another along a trajectory which makes the integral $\int p(x,y,z)\,ds$, where $p(x,y,z) = \sqrt{2m(E - V(x,y,z))}$ is the magnitude of the momentum, stationary (for further details, see Chapter VIII, *Lagrangian and Hamiltonian Mechanics*).
(a) Consider projectile motion in the (x,y) plane with x horizontal and y vertical, and with potential $V = mgy$ where g is the (constant) gravitational field. Write down the Euler-Lagrange equation which results from Jacobi's principle, and integrate to obtain the equation for the trajectories.
(Ans. $y - y_0 = (x - x_0)\tan\alpha - \dfrac{(x - x_0)^2}{4h\cos^2\alpha}$ where (x_0, y_0) is the start point, $E = mg(h + y_0)$ is the energy, and α is the angle of launch)
(b) Sketch the family of trajectories which start at $(0,0)$ with fixed energy $E = mgh$ but arbitrary angle α of launch. Show that if the end point lies within the envelope $y = h - (x^2/4h)$ of the family of trajectories, two trajectories connect the start and end points, whereas if the end point lies outside the envelope, no trajectories connect the start and end points.

Solution

(a) According to Jacobi's principle, a projectile of mass m and energy E, which moves in the xy-plane in the earth's gravitational field $-g\mathbf{j}$, follows a trajectory which makes the integral

$$\int p\,ds = \int_{x_0}^{x_1} \sqrt{2m(E-mgy)}\sqrt{1+(dy/dx)^2}\,dx$$

stationary. This is a variational principle with integrand

$$F = \sqrt{2m(E-mgy)}\sqrt{1+(dy/dx)^2}\,.$$

Since F does not depend explicitly on the independent variable x, for an actual trajectory the quantity

$$H = \frac{\partial F}{\partial(dy/dx)}\frac{dy}{dx} - F = -\sqrt{\frac{2m(E-mgy)}{1+(dy/dx)^2}}$$

is constant. The value of this constant is determined by the initial conditions. We set $E = mg(h+y_0)$ where mgh is the initial kinetic energy and note that $(dy/dx)_0 = \tan\alpha$ where α is the angle of launch. The constant H is thus

$$H = -\sqrt{\frac{2m(E-mgy_0)}{1+(dy/dx)_0^2}} = -\sqrt{2m^2gh\cos^2\alpha}\,,$$

and the preceding equation becomes

$$\frac{h-(y-y_0)}{1+(dy/dx)^2} = h\cos^2\alpha.$$

Solving for dy/dx, we obtain

$$\left(\frac{dy}{dx}\right)^2 = \tan^2\alpha\left[1-\frac{y-y_0}{h\sin^2\alpha}\right].$$

Rearranging this and integrating then gives the equation for a trajectory

$$(x-x_0)\tan\alpha = \int_{y_0}^{y}\frac{dy}{\sqrt{1-(y-y_0)/(h\sin^2\alpha)}} = 2h\sin^2\alpha\left[1-\sqrt{1-\frac{y-y_0}{h\sin^2\alpha}}\right].$$

We rewrite this in the form

$$y - y_0 = (x - x_0)\tan\alpha - \frac{(x - x_0)^2}{4h\cos^2\alpha}$$

which we recognize as the equation of a parabola.

(b) The trajectories

$$y = x\tan\alpha - \frac{x^2}{4h\cos^2\alpha}$$

which start at $(0,0)$ are, for fixed h and variable α, a family of parabolas with range

$$R = 4h\sin\alpha\cos\alpha = 2h\sin 2\alpha$$

and maximum height $h\sin^2\alpha$. Clearly, for $\alpha = 0$ (horizontal launch) both the range and maximum height are zero. As α is increased, both the range and maximum height at first increase, until at $\alpha = \pi/4$ a maximum range of $2h$ is achieved; the maximum height of this trajectory is $h/2$. As α is increased still further, the range now decreases while the maximum height continues to increase, until at $\alpha = \pi/2$ (vertical launch) the range is again zero and the maximum height is h. There are two possible trajectories for any range less than $2h$ (Fig. 1).

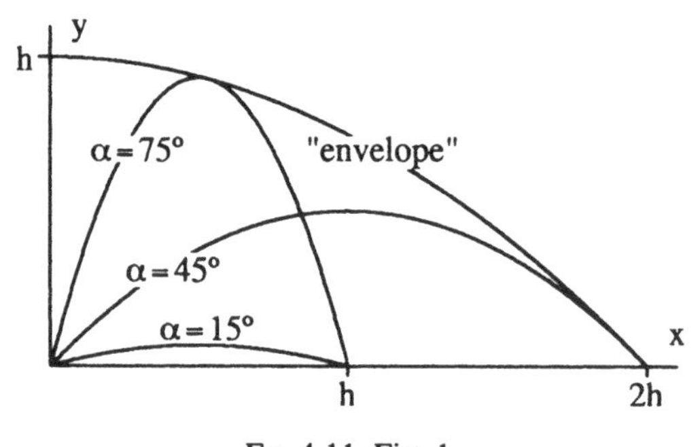

Ex. 4.11, Fig. 1

Indeed, for any end point within the envelope of this family of parabolas there are two possible trajectories. Note that the equation for the trajectories can be written

$$y = x\tan\alpha - \frac{x^2}{4h}(1 + \tan^2\alpha).$$

This is, for given end point (x, y), a quadratic equation for $\tan\alpha$, the solutions of which are

$$\tan\alpha = \frac{2h}{x}\left[1 \pm \sqrt{1 - \frac{y}{h} - \frac{x^2}{4h^2}}\right].$$

For $y < h - x^2/4h$ there are two real solutions for $\tan\alpha$ (and hence for α); for $y = h - x^2/4h$ there one real solution; and for $y > h - x^2/4h$ there are no real solutions. The curve $y = h - x^2/4h$ is the envelope of the family of trajectories which start at the origin with fixed energy $E = mgh$ but arbitrary angle of launch.

Exercise 4.12

A particle moves vertically in the uniform gravitational field g near the surface of the earth. The Lagrangian is

$$L = \tfrac{1}{2}m\dot{z}^2 - mgz.$$

Suppose that at time 0 the particle is at $z = 0$ and at time t_1 it is at $z = z_1$. For any motion $z(t)$, actual or virtual, between these two points the action is

$$S[z(t)] = \int_0^{t_1} L(z,\dot{z})\,dt.$$

Pretend you don't know what the actual motion is. You might then guess that it can be adequately represented by the first three terms in a power series in t,

$$z = z_0 + v_0 t + \tfrac{1}{2}at^2,$$

where z_0 and v_0 are chosen so that $z(t)$ passes through the end points, and a is an adjustable parameter. Evaluate S for this form of $z(t)$ and note the dependence of S on a. For what value of a is S a minimum?

Solution

The action along the path

$$z = z_0 + v_0 t + \tfrac{1}{2}at^2$$

from $t = 0$ to $t = t_1$, for a particle which moves vertically in a uniform gravitational field g, is

$$S = \int_0^{t_1} \left[\tfrac{1}{2}m(v_0 + at)^2 - mg(z_0 + v_0 t + \tfrac{1}{2}at^2)\right]dt$$
$$= \tfrac{1}{2}m(v_0^2 t_1 + v_0 a t_1^2 + \tfrac{1}{3}a^2 t_1^3) - mg(z_0 t_1 + \tfrac{1}{2}v_0 t_1^2 + \tfrac{1}{6}a t_1^3).$$

For paths which start at $(t = 0, z = 0)$ and end at $(t = t_1, z = z_1)$ we have

$$0 = z_0 \quad \text{and} \quad z_1 = v_0 t_1 + \tfrac{1}{2}a t_1^2,$$

so the initial velocity is

$$v_0 = \frac{z_1}{t_1} - \frac{1}{2}a t_1.$$

Using these relations to eliminate z_0 and v_0 from our expression for S, we find

$$S = \frac{m z_1^2}{2t_1} - \frac{m g z_1 t_1}{2} + \frac{1}{24}m(a^2 + 2ag)t_1^3.$$

The action S for these paths between the given end points is a quadratic function of the adjustable parameter "a." Minimum action, and the "best" path of the form chosen, occurs for $a = -g$. Indeed, this gives the actual path, which happens to be a special case of paths of the form chosen.

CHAPTER V

INVARIANCE TRANSFORMATIONS AND CONSTANTS OF THE MOTION

Exercise 5.01

Show that a function of $q(t)$, $\dot{q}(t)$, and t satisfies Lagrange's equations identically (independent of $q_a(t)$) if, and only if, it is the total time derivative $d\Lambda/dt$ of some function $\Lambda(q(t),t)$.

Solution

The "if": let

$$F = \frac{d\Lambda(q,t)}{dt} = \sum_{b=1}^{f} \frac{\partial \Lambda}{\partial q_b}\dot{q}_b + \frac{\partial \Lambda}{\partial t}.$$

Now

$$\frac{\partial F}{\partial \dot{q}_a} = \frac{\partial \Lambda}{\partial q_a}, \quad \frac{d}{dt}\left(\frac{\partial F}{\partial \dot{q}_a}\right) = \sum_{b=1}^{f} \frac{\partial^2 \Lambda}{\partial q_b \partial q_a}\dot{q}_b + \frac{\partial^2 \Lambda}{\partial t \partial q_a}, \quad \text{and} \quad \frac{\partial F}{\partial q_a} = \sum_{b=1}^{f} \frac{\partial^2 \Lambda}{\partial q_a \partial q_b}\dot{q}_b + \frac{\partial^2 \Lambda}{\partial q_a \partial t}.$$

Comparing, we see that F satisfies Lagrange's equations,

$$\frac{d}{dt}\left(\frac{\partial F}{\partial \dot{q}_a}\right) = \frac{\partial F}{\partial q_a},$$

identically (independent of the form of Λ and of the specific time dependence of q).

The "only if": conversely, suppose that F is some function of q, $\dot{q}$, and t, which satisfies Lagrange's equations identically. These equations can be written

$$\sum_{b=1}^{f} \frac{\partial^2 F}{\partial \dot{q}_b \partial \dot{q}_a}\ddot{q}_b + \sum_{b=1}^{f} \frac{\partial^2 F}{\partial q_b \partial \dot{q}_a}\dot{q}_b + \frac{\partial^2 F}{\partial t \partial \dot{q}_a} = \frac{\partial F}{\partial q_a}.$$

The only place the $\ddot{q}$'s appear is in the first term on the left. Since we want the equations to be satisfied independent of q, the coefficients of each of the $\ddot{q}$'s must be zero,

$$\frac{\partial^2 F}{\partial \dot{q}_b \partial \dot{q}_a} = 0.$$

This means that F is a linear function of the $\dot{q}$'s,

$$F(q,\dot{q},t) = \sum_{b=1}^{f} f_b(q,t)\dot{q}_b + g(q,t).$$

Now

$$\frac{\partial F}{\partial \dot{q}_a} = f_a, \quad \frac{d}{dt}\left(\frac{\partial F}{\partial \dot{q}_a}\right) = \sum_{b=1}^{f} \frac{\partial f_a}{\partial q_b}\dot{q}_b + \frac{\partial f_a}{\partial t}, \quad \text{and} \quad \frac{\partial F}{\partial q_a} = \sum_{b=1}^{f} \frac{\partial f_b}{\partial q_a}\dot{q}_b + \frac{\partial g}{\partial q_a}.$$

Substituting these into Lagrange's equations, we find

$$\sum_{b=1}^{f}\left(\frac{\partial f_a}{\partial q_b} - \frac{\partial f_b}{\partial q_a}\right)\dot{q}_b + \frac{\partial f_a}{\partial t} = \frac{\partial g}{\partial q_a}.$$

The only place the $\dot{q}$'s appear is in the first term on the left. Since we want the equations to be satisfied independent of q, the coefficients of each of the $\dot{q}$'s must be zero,

$$\frac{\partial f_a}{\partial q_b} = \frac{\partial f_b}{\partial q_a}.$$

These are the conditions for the f's to be expressible in the form

$$f_a(q,t) = \frac{\partial \Lambda(q,t)}{\partial q_a}$$

where $\Lambda(q,t)$ is some function of the coordinates q and the time t. Substituting this into Lagrange's equations, we find

$$\frac{\partial^2 \Lambda}{\partial t \partial q_a} = \frac{\partial g}{\partial q_a},$$

and so

$$g = \frac{\partial \Lambda}{\partial t}.$$

Thus, in order to satisfy Lagrange's equations identically, a function $F(q,\dot{q},t)$ must have the form

$$F = \sum_{b=1}^{f} \frac{\partial \Lambda}{\partial q_b}\dot{q}_b + \frac{\partial \Lambda}{\partial t} = \frac{d\Lambda(q,t)}{dt}.$$

Exercise 5.02*

The motion of a particle of mass m which moves vertically in the uniform gravitational field g near the surface of the earth can be described by an action principle with Lagrangian

$$L = \tfrac{1}{2}m\dot{z}^2 - mgz.$$

(a) Show that the action principle is invariant under the transformation $z' = z + \alpha$ where α is any constant, and find the associated constant of the motion.
(b) Show that the action principle is invariant under the transformation $z' = z + \beta t$ where β is any constant, and find the associated constant of the motion.

Solution

(a) The new Lagrangian $L'(z', \dot{z}', t)$ for the new variable z' is obtained by transforming variables, setting

$$z = z' - \alpha \quad \text{and} \quad \dot{z} = \dot{z}'$$

in the old Lagrangian $L(z, \dot{z}, t)$. We have

$$\begin{aligned} L'(z', \dot{z}') &= L(z, \dot{z}) \\ &= \tfrac{1}{2}m\dot{z}^2 - mgz \\ &= \tfrac{1}{2}m\dot{z}'^2 - mg(z' - \alpha) \\ &= \tfrac{1}{2}m\dot{z}'^2 - mgz' + mg\alpha \\ &= L(z', \dot{z}') + d\Lambda/dt \end{aligned}$$

where $\Lambda = mgt\alpha$. Thus the action principle and system are invariant under this transformation.

The corresponding infinitesimal invariance transformation is obtained by replacing α by $\delta\alpha$ and setting

$$\delta z = z' - z = \delta\alpha \quad \text{and} \quad \delta\Lambda = mgt\,\delta\alpha.$$

The associated constant of the motion is

$$\begin{aligned} \frac{\partial L}{\partial \dot{z}}\delta z + \delta\Lambda &= m\dot{z}\,\delta\alpha + mgt\,\delta\alpha \\ &= m(\dot{z} + gt)\,\delta\alpha\,. \end{aligned}$$

The constant $m(\dot{z} + gt)$ equals mv_0 where v_0 is the initial velocity of the particle.

(b) The new Lagrangian $L'(z',\dot{z}',t)$ for the new variable z' is obtained by transforming variables, setting

$$z = z' - \beta t \quad \text{and} \quad \dot{z} = \dot{z}' - \beta$$

in the old Lagrangian $L(z,\dot{z},t)$. We have

$$\begin{aligned} L'(z',\dot{z}') &= L(z,\dot{z}) \\ &= \tfrac{1}{2}m\dot{z}^2 - mgz \\ &= \tfrac{1}{2}m(\dot{z}' - \beta)^2 - mg(z' - \beta t) \\ &= \tfrac{1}{2}m\dot{z}'^2 - mgz' - m\dot{z}'\beta + \tfrac{1}{2}m\beta^2 + mgt\beta \\ &= L(z',\dot{z}') + d\Lambda/dt \end{aligned}$$

where $\Lambda = -m(z' - \frac{1}{2}gt^2)\beta + \frac{1}{2}m\beta^2 t$. Thus the action principle and system are invariant under this transformation.

The corresponding infinitesimal invariance transformation is obtained by replacing β by $\delta\beta$ and setting

$$\delta z = z' - z = t\,\delta\beta \quad \text{and} \quad \delta\Lambda = -m(z - \tfrac{1}{2}gt^2)\delta\beta.$$

The associated constant of the motion is

$$\begin{aligned} \frac{\partial L}{\partial \dot{z}}\delta z + \delta\Lambda &= m\dot{z}t\,\delta\beta - m(z - \tfrac{1}{2}gt^2)\delta\beta \\ &= m(\dot{z}t - z + \tfrac{1}{2}gt^2)\delta\beta\,. \end{aligned}$$

The constant $m(\dot{z}t - z + \frac{1}{2}gt^2)$ equals $-mz_0$ where z_0 is the initial position of the particle.

The expressions

$$\dot{z} + gt = v_0 \quad \text{and} \quad \dot{z}t - z + \tfrac{1}{2}gt^2 = -z_0$$

for the two constants of the motion can be inverted to obtain the velocity and position of the particle as functions of time,

$$\dot{z} = v_0 - gt \quad \text{and} \quad z = z_0 + v_0 t - \tfrac{1}{2}gt^2,$$

thus solving the equations of motion.

Exercise 5.03

The motion of a simple harmonic oscillator is described by an action principle with Lagrangian

$$L = \tfrac{1}{2}m\dot{x}^2 - \tfrac{1}{2}m\omega^2 x^2$$

(a) Show that the action principle is invariant under the two-parameter (A, B) family of transformations

$$x' = x + A\sin\omega t + B\cos\omega t.$$

(b) Find the two independent constants of the motion associated with the infinitesimal transformation, and identify them physically.
(c) Use the results of (b) to write down the general solution to the equation of motion.

Solution

(a) The new Lagrangian $L'(x', \dot{x}', t)$ for the new variable x' is obtained by transforming variables, setting

$$x = x' - A\sin\omega t - B\cos\omega t \quad \text{and} \quad \dot{x} = \dot{x}' - \omega A\cos\omega t + \omega B\sin\omega t$$

in the old Lagrangian $L(x, \dot{x})$. We have

$$\begin{aligned}
L'(x', \dot{x}', t) &= L(x, \dot{x}) \\
&= \tfrac{1}{2}m\dot{x}^2 - \tfrac{1}{2}m\omega^2 x^2 \\
&= \tfrac{1}{2}m(\dot{x}' - \omega A\cos\omega t + \omega B\sin\omega t)^2 - \tfrac{1}{2}m\omega^2(x' - A\sin\omega t - B\cos\omega t)^2 \\
&= \tfrac{1}{2}m\dot{x}'^2 + m\dot{x}'(-\omega A\cos\omega t + \omega B\sin\omega t) + \tfrac{1}{2}m(-\omega A\cos\omega t + \omega B\sin\omega t)^2 \\
&\quad - \tfrac{1}{2}m\omega^2 x'^2 - m\omega^2 x'(-A\sin\omega t - B\cos\omega t) - \tfrac{1}{2}m\omega^2(-A\sin\omega t - B\cos\omega t)^2 \\
&= L(x', \dot{x}') + d\Lambda/dt
\end{aligned}$$

where

$$\Lambda = m\omega x'(-A\cos\omega t + B\sin\omega t) + \tfrac{1}{4}m\omega(A^2 - B^2)\sin 2\omega t.$$

Thus the action principle and system are invariant under this transformation.

(b) The corresponding infinitesimal invariance transformation is

$$\delta x = \delta A\sin\omega t + \delta B\cos\omega t \quad \text{and} \quad \delta\Lambda = m\omega x(-\delta A\cos\omega t + \delta B\sin\omega t).$$

The associated constant of the motion is

$$\frac{\partial L}{\partial \dot{x}}\delta x + \delta\Lambda = m\dot{x}(\delta A \sin\omega t + \delta B \cos\omega t) + m\omega x(-\delta A \cos\omega t + \delta B \sin\omega t)$$
$$= m(\dot{x}\sin\omega t - \omega x \cos\omega t)\delta A + m(\dot{x}\cos\omega t + \omega x \sin\omega t)\delta B .$$

Since δA and δB are independent, the coefficient of δA and the coefficient of δB are separately constants of motion. We can identify these constants by looking at $t = 0$. At this time the coefficient of δA becomes $-m\omega x_0$ where x_0 is the initial position, and the coefficient of δB becomes mv_0 where v_0 is the initial velocity,

$$m(\dot{x}\sin\omega t - \omega x \cos\omega t) = -m\omega x_0 ,$$
$$m(\dot{x}\cos\omega t + \omega x \sin\omega t) = mv_0 .$$

(c) These expressions for the two constants of the motion can be inverted to obtain the position and velocity as functions of time,

$$x = x_0 \cos\omega t + (v_0/\omega)\sin\omega t,$$
$$\dot{x} = -\omega x_0 \sin\omega t + v_0 \cos\omega t ,$$

thus solving the equations of motion.

Exercise 5.04

The Lagrangian for a particle of mass m and charge e moving in a uniform magnetic field which points in the z-direction is (see Exercise 3.14)

$$L = \tfrac{1}{2}m(\dot{x}^2 + \dot{y}^2 + \dot{z}^2) + (eB/2c)(x\dot{y} - y\dot{x}).$$

(a) Show that the system is invariant under spatial displacement (in any direction) and find the associated constants of the motion.
(b) Show that the system is invariant under rotation about the z-axis and find the associated constant of the motion.

Solution

(a) First consider spatial displacement. The new Lagrangian $L'(q',\dot{q}')$ for the new variables q' is obtained by transforming the variables, setting

$$x = x' - a_x , \qquad \dot{x} = \dot{x}',$$
$$y = y' - a_y , \quad \text{and} \quad \dot{y} = \dot{y}',$$
$$z = z' - a_z , \qquad \dot{z} = \dot{z}',$$

in the old Lagrangian $L(q,\dot{q})$. We have

$$\begin{aligned} L'(q',\dot{q}') &= L(q,\dot{q}) \\ &= \tfrac{1}{2}m(\dot{x}^2 + \dot{y}^2 + \dot{z}^2) + (eB/2c)(x\dot{y} - y\dot{x}) \\ &= \tfrac{1}{2}m(\dot{x}'^2 + \dot{y}'^2 + \dot{z}'^2) + (eB/2c)((x' - a_x)\dot{y}' - (y' - a_y)\dot{x}') \\ &= \tfrac{1}{2}m(\dot{x}'^2 + \dot{y}'^2 + \dot{z}'^2) + (eB/2c)(x'\dot{y}' - y'\dot{x}') + (eB/2c)(-a_x\dot{y}' + a_y\dot{x}') \\ &= L(q',\dot{q}') + d\Lambda/dt \end{aligned}$$

where $\Lambda = (eB/2c)(-a_x y' + a_y x')$. Thus the system is invariant under spatial displacement.

The corresponding infinitesimal transformation is

$$\delta x = \delta a_x, \quad \delta y = \delta a_y, \quad \delta z = \delta a_z, \quad \text{and} \quad \delta\Lambda = (eB/2c)(-y\,\delta a_x + x\,\delta a_y).$$

The associated constant of the motion is

$$\begin{aligned} \sum_{a=1}^{f} \frac{\partial L}{\partial \dot{q}_a}\delta q_a + \delta\Lambda &= \left(m\dot{x} - \frac{eB}{2c}y\right)\delta a_x + \left(m\dot{y} + \frac{eB}{2c}x\right)\delta a_y + m\dot{z}\delta a_z + \frac{eB}{2c}(-y\,\delta a_x + x\,\delta a_y) \\ &= \left(m\dot{x} - \frac{eB}{c}y\right)\delta a_x + \left(m\dot{y} + \frac{eB}{c}x\right)\delta a_y + m\dot{z}\delta a_z\,. \end{aligned}$$

Since the displacements in the x-, y-, and z-directions are independent, the coefficients of δa_x, δa_y, and δa_z,

$$\mu = m\dot{x} - (eB/c)y, \quad \nu = m\dot{y} + (eB/c)x, \quad \text{and} \quad p_{\|} = m\dot{z},$$

are separately constants of the motion.

(b) Now consider rotation about the z-axis. The new Lagrangian $L'(q',\dot{q}')$ for the new variables q' is obtained by transforming the variables, setting

$$\begin{aligned} x &= x'\cos\theta + y'\sin\theta, & \dot{x} &= \dot{x}'\cos\theta + \dot{y}'\sin\theta, \\ y &= -x'\sin\theta + y'\cos\theta, \quad \text{and} & \dot{y} &= -\dot{x}'\sin\theta + \dot{y}'\cos\theta, \\ z &= z', & \dot{z} &= \dot{z}', \end{aligned}$$

in the old Lagrangian $L(q,\dot{q})$. We have

$$\begin{aligned} L'(q',\dot{q}') &= L(q,\dot{q}) \\ &= \tfrac{1}{2}m(\dot{x}^2 + \dot{y}^2 + \dot{z}^2) + (eB/2c)(x\dot{y} - y\dot{x}) \\ &= \tfrac{1}{2}m[(\dot{x}'\cos\theta + \dot{y}'\sin\theta)^2 + (-\dot{x}'\sin\theta + \dot{y}'\cos\theta)^2 + \dot{z}'^2] \\ &\quad +(eB/2c)[(x'\cos\theta + y'\sin\theta)(-\dot{x}'\sin\theta + \dot{y}'\cos\theta) \\ &\quad -(-x'\sin\theta + y'\cos\theta)(\dot{x}'\cos\theta + \dot{y}'\sin\theta)] \\ &= \tfrac{1}{2}m(\dot{x}'^2 + \dot{y}'^2 + \dot{z}'^2) + (eB/2c)(x'\dot{y}' - y'\dot{x}') \\ &= L(q',\dot{q}')\,. \end{aligned}$$

Thus the system is invariant under rotation about the z-axis.

The corresponding infinitesimal transformation is

$$\delta x = -y\,\delta\theta, \quad \delta y = x\,\delta\theta, \quad \delta z = 0, \quad \text{and} \quad \delta\Lambda = 0.$$

The associated constant of the motion is

$$\begin{aligned}\sum_{a=1}^{f} \frac{\partial L}{\partial \dot q_a}\delta q_a + \delta\Lambda &= \left(m\dot x - \frac{eB}{2c}y\right)(-y\,\delta\theta) + \left(m\dot y + \frac{eB}{2c}x\right)(x\,\delta\theta) \\ &= \left[m(x\dot y - y\dot x) + (eB/2c)(x^2+y^2)\right]\delta\theta\,.\end{aligned}$$

Let

$$\lambda = m(x\dot y - y\dot x) + (eB/2c)(x^2+y^2)$$

denote this constant of the motion.

It is of interest to use the constants of the motion μ and ν, previously obtained, to eliminate $m\dot x = \mu + (eB/c)y$ and $m\dot y = \nu - (eB/c)x$ from the expression for λ. We have

$$\lambda = x(\nu - (eB/c)x) - y(\mu + (eB/c)y) + (eB/2c)(x^2+y^2)\,.$$

This equation can be rewritten

$$\left(x - \frac{c\nu}{eB}\right)^2 + \left(y + \frac{c\mu}{eB}\right)^2 = \left(\frac{c}{eB}\right)^2\left(\mu^2 + \nu^2 - 2\frac{eB}{c}\lambda\right)$$

and shows that the orbit lies on a circular cylinder with axis parallel to the z-axis, with center at $x = c\nu/eB$, $y = -c\mu/eB$, and with radius $R = (c/eB)\sqrt{\mu^2 + \nu^2 - 2(eB/c)\lambda}$. This identifies, physically, the constants μ, ν, and λ.

Since the Lagrangian does not depend explicitly on the time, there is another constant of the motion, the energy (Hamiltonian)

$$\begin{aligned}E &= \sum_{a=1}^{f} \frac{\partial L}{\partial \dot q_a}\dot q_a - L \\ &= \left(m\dot x - \frac{eB}{2c}y\right)\dot x + \left(m\dot y + \frac{eB}{2c}x\right)\dot y + m\dot z^2 - \frac{1}{2}m(\dot x^2 + \dot y^2 + \dot z^2) - \frac{eB}{2c}(x\dot y - y\dot x) \\ &= \frac{1}{2}m(\dot x^2 + \dot y^2 + \dot z^2)\,.\end{aligned}$$

However, the energy is not independent of the constants already obtained. Indeed, we have

$$E = \frac{1}{2m}\left[\left(\mu + \frac{eB}{c}y\right)^2 + \left(\nu - \frac{eB}{c}x\right)^2 + p_\parallel^2\right]$$
$$= \frac{1}{2m}\left(\mu^2 + \nu^2 + p_\parallel^2\right) + \frac{eB}{mc}(\mu y - \nu x) + \frac{1}{2m}\left(\frac{eB}{c}\right)^2\left(x^2 + y^2\right)$$
$$= \frac{1}{2m}\left(\mu^2 + \nu^2 + p_\parallel^2\right) - \frac{eB}{mc}\lambda\,.$$

The radius R of the circular cylinder on which the orbit lies can be expressed simply in terms of the energy E,

$$R = (c/eB)\sqrt{2mE - p_\parallel^2} = v_\perp/\omega.$$

In the second equality $v_\perp = \sqrt{\dot{x}^2 + \dot{y}^2}$ is the velocity perpendicular to the magnetic field and $\omega = eB/mc$ is the cyclotron frequency.

CHAPTER VI

HAMILTON'S EQUATIONS

Exercise 6.01

A system with one degree of freedom has a Hamiltonian

$$H(q,p) = \frac{p^2}{2m} + A(q)p + B(q)$$

where A and B are certain functions of the coordinate q, and p is the momentum conjugate to q.
(a) Find the velocity $\dot{q}$.
(b) Find the Lagrangian $L(q,\dot{q})$ (note variables).

Solution

(a) The velocity $\dot{q}$ is given by the first of Hamilton's equations,

$$\dot{q} = \frac{\partial H}{\partial p} = \frac{p}{m} + A(q).$$

(b) The Lagrangian is then given by

$$L(q,\dot{q}) = p(q,\dot{q})\dot{q} - H(q,p(q,\dot{q})) = \tfrac{1}{2}m(\dot{q} - A)^2 - B.$$

This illustrates how one can go from Hamiltonian mechanics back to Lagrangian mechanics.

Exercise 6.02

We have seen (Exercise 5.01) that two Lagrangians L' and L which differ by the total time derivative $d\Lambda/dt$ of some function $\Lambda(q,t)$,

$$L' = L + d\Lambda/dt,$$

are equivalent, leading to the same Lagrange's equations of motion.
(a) What is the relation between the generalized momenta p' and p which these two Lagrangians yield?
(b) What is the relation between the Hamiltonians H' and H which these two Lagrangians yield?
(c) Show explicitly that Hamilton's equations of motion in the primed quantities are equivalent to those in the unprimed quantities.

Solution

(a) and (b) Starting with a Lagrangian L, we introduce a new Lagrangian L′ by a transformation of the form

$$L' = L + \frac{d\Lambda}{dt} = L + \sum_{a=1}^{f} \frac{\partial \Lambda}{\partial q_a}\dot{q}_a + \frac{\partial \Lambda}{\partial t}.$$

The new generalized momentum p′ is

$$p'_a = \frac{\partial L'}{\partial \dot{q}_a} = \frac{\partial L}{\partial \dot{q}_a} + \frac{\partial \Lambda}{\partial q_a} = p_a + \frac{\partial \Lambda}{\partial q_a},$$

and the new Hamiltonian H′ is

$$H' = \sum_{a=1}^{f} p'_a \dot{q}_a - L' = \sum_{a=1}^{f}\left(p_a + \frac{\partial \Lambda}{\partial q_a}\right)\dot{q}_a - L - \sum_{a=1}^{f} \frac{\partial \Lambda}{\partial q_a}\dot{q}_a - \frac{\partial \Lambda}{\partial t} = H\left(q, p' - \frac{\partial \Lambda}{\partial q}\right) - \frac{\partial \Lambda}{\partial t}.$$

This transformation is sometimes called a "gauge transformation." Compare the behavior, under electromagnetic gauge transformations, of the generalized momentum of a charged particle in an electromagnetic field.

(c) Hamilton's equations for the new variables are

$$\frac{dq_a}{dt} = \frac{\partial H'}{\partial p'_a} \quad \text{and} \quad \frac{dp'_a}{dt} = -\left(\frac{\partial H'}{\partial q_a}\right)_{p'}.$$

Looking at our expressions for H′ and p′, we see that the right hand side of the first of these equations is equal to $\partial H/\partial p_a$, so this equation becomes

$$\frac{dq_a}{dt} = \frac{\partial H}{\partial p_a}$$

which is the first Hamilton equation for the old variables. The second Hamilton equation for the new variables becomes

$$\frac{d}{dt}\left(p_a + \frac{\partial \Lambda}{\partial q_a}\right) = -\left[\frac{\partial}{\partial q_a}\left(H - \frac{\partial \Lambda}{\partial t}\right)\right]_{p'}.$$

Expanding both sides of this equation, we find

$$\frac{dp_a}{dt} + \sum_{b=1}^{f} \frac{\partial^2 \Lambda}{\partial q_b \partial q_a}\frac{dq_b}{dt} + \frac{\partial^2 \Lambda}{\partial t \partial q_a} = -\left(\frac{\partial H}{\partial q_a}\right)_p + \sum_{b=1}^{f}\left(\frac{\partial H}{\partial p_b}\right)_q \frac{\partial^2 \Lambda}{\partial q_a \partial q_b} + \frac{\partial^2 \Lambda}{\partial q_a \partial t}$$

which, in view of the first Hamilton equation for the old variables, reduces to the second Hamilton equation

$$\frac{dp_a}{dt} = -\left(\frac{\partial H}{\partial q_a}\right)_p$$

for the old variables. Hamilton's equations for the new variables are thus equivalent to those for the old variables.

Exercise 6.03

A particle of mass m moves in a central force field with potential $V(r)$. The Lagrangian in terms of spherical polar coordinates (r,θ,ϕ) is

$$L = \tfrac{1}{2}m\left(\dot{r}^2 + r^2\dot{\theta}^2 + r^2\sin^2\theta\,\dot{\phi}^2\right) - V(r).$$

(a) Find the momenta (p_r, p_θ, p_ϕ) conjugate to (r,θ,ϕ).
(b) Find the Hamiltonian $H(r,\theta,\phi,p_r,p_\theta,p_\phi)$.
(c) Write down the explicit Hamilton's equations of motion.

Solution

(a) The momenta (p_r, p_θ, p_ϕ) conjugate to (r,θ,ϕ) are

$$p_r = \frac{\partial L}{\partial \dot{r}} = m\dot{r}, \quad p_\theta = \frac{\partial L}{\partial \dot{\theta}} = mr^2\dot{\theta}, \quad \text{and} \quad p_\phi = \frac{\partial L}{\partial \dot{\phi}} = mr^2\sin^2\theta\,\dot{\phi}.$$

(b) The Hamiltonian $H(r,\theta,\phi,p_r,p_\theta,p_\phi)$ is

$$\begin{aligned} H &= p_r\dot{r} + p_\theta\dot{\theta} + p_\phi\dot{\phi} - L \\ &= \frac{p_r^2}{2m} + \frac{p_\theta^2}{2mr^2} + \frac{p_\phi^2}{2mr^2\sin^2\theta} + V(r). \end{aligned}$$

(c) Hamilton's equations of motion are

$$\frac{dr}{dt} = \frac{\partial H}{\partial p_r} = \frac{p_r}{m}, \qquad \frac{dp_r}{dt} = -\frac{\partial H}{\partial r} = \frac{p_\theta^2}{mr^3} + \frac{p_\phi^2}{mr^3\sin^2\theta} - \frac{\partial V(r)}{\partial r},$$

$$\frac{d\theta}{dt} = \frac{\partial H}{\partial p_\theta} = \frac{p_\theta}{mr^2}, \qquad \frac{dp_\theta}{dt} = -\frac{\partial H}{\partial \theta} = \frac{p_\phi^2\cos\theta}{mr^2\sin^3\theta},$$

$$\frac{d\phi}{dt} = \frac{\partial H}{\partial p_\phi} = \frac{p_\phi}{mr^2\sin^2\theta}, \qquad \frac{dp_\phi}{dt} = -\frac{\partial H}{\partial \phi} = 0.$$

Exercise 6.04

The Lagrangian for a free particle in terms of paraboloidal coordinates (ξ,η,ϕ) is (see Exercise 3.09)

$$L = \tfrac{1}{2}m\left(\xi^2+\eta^2\right)\left(\dot{\xi}^2+\dot{\eta}^2\right)+\tfrac{1}{2}m\xi^2\eta^2\dot{\phi}^2.$$

(a) Find the momenta conjugate to (ξ,η,ϕ).
(b) Find the Hamiltonian.

Solution

(a) The momenta (p_ξ, p_η, p_ϕ) conjugate to (ξ,η,ϕ) are

$$p_\xi = \frac{\partial L}{\partial\dot{\xi}} = m(\xi^2+\eta^2)\dot{\xi}, \quad p_\eta = \frac{\partial L}{\partial\dot{\eta}} = m(\xi^2+\eta^2)\dot{\eta}, \quad \text{and} \quad p_\phi = \frac{\partial L}{\partial\dot{\phi}} = m\xi^2\eta^2\,\dot{\phi}.$$

(b) The Hamiltonian is

$$\begin{aligned} H &= p_\xi\dot{\xi}+p_\eta\dot{\eta}+p_\phi\dot{\phi}-L \\ &= \frac{1}{2m}\left(\frac{p_\xi^2+p_\eta^2}{\xi^2+\eta^2}+\frac{p_\phi^2}{\xi^2\eta^2}\right). \end{aligned}$$

Exercise 6.05*

The Lagrangian for a free particle of mass m, referred to cartesian coordinates (x',y',z') which are rotating about an inertial z-axis with angular velocity ω, is (see Exercise 3.13)

$$L' = \tfrac{1}{2}m\left[(\dot{x}'^2+\dot{y}'^2+\dot{z}'^2)+2\omega(x'\dot{y}'-y'\dot{x}')+\omega^2(x'^2+y'^2)\right].$$

(a) Find the momenta (p_x', p_y', p_z') conjugate to (x',y',z').
(b) Find the Hamiltonian $H'(x',y',z',p_x',p_y',p_z')$.
(Ans. $H' = \frac{1}{2m}\left(p_x'^2+p_y'^2+p_z'^2\right)-\omega\left(x'p_y'-y'p_x'\right)$)

Solution

(a) The momenta (p_x', p_y', p_z') conjugate to (x',y',z') are

$$p_x' = \frac{\partial L'}{\partial\dot{x}'} = m(\dot{x}'-\omega y'), \quad p_y' = \frac{\partial L'}{\partial\dot{y}'} = m(\dot{y}'+\omega x'), \quad \text{and} \quad p_z' = \frac{\partial L'}{\partial\dot{z}'} = m\dot{z}'.$$

Note that (see Exercise 3.13)

$$p'_x = m\big(\dot{x}\cos\theta(t) + \dot{y}\sin\theta(t)\big), \quad p'_y = m\big(-\dot{x}\sin\theta(t) + \dot{y}\cos\theta(t)\big), \quad \text{and} \quad p'_z = m\dot{z},$$

so the (canonical) momentum vector $\mathbf{p}'$ in the rotating frame is

$$\begin{aligned}
\mathbf{p}' &= p'_x\mathbf{i}' + p'_y\mathbf{j}' + p'_z\mathbf{k}' \\
&= m\dot{x}\big(\mathbf{i}'\cos\theta(t) - \mathbf{j}'\sin\theta(t)\big) + m\dot{y}\big(\mathbf{i}'\sin\theta(t) + \mathbf{j}'\cos\theta(t)\big) + m\dot{z}\mathbf{k}' \\
&= m\dot{x}\mathbf{i} + m\dot{y}\mathbf{j} + m\dot{z}\mathbf{k} = \mathbf{p}\,.
\end{aligned}$$

It equals the (canonical and kinematic) momentum vector $\mathbf{p} = m\dot{\mathbf{r}}$ in the inertial frame.

(b) The Hamiltonian $H'(x', y', z', p'_x, p'_y, p'_z)$ is

$$\begin{aligned}
H' &= p'_x\dot{x}' + p'_y\dot{y}' + p'_z\dot{z}' - L' \\
&= \frac{1}{2m}\left(p'^2_x + p'^2_y + p'^2_z\right) - \omega\left(x'p'_y - y'p'_x\right).
\end{aligned}$$

The first term in H' is $|\mathbf{p}'|^2/2m$ and equals the Hamiltonian $H = |\mathbf{p}|^2/2m$ in the inertial frame. The second term is

$$-\omega\left(x'p'_y - y'p'_x\right) = -\omega\left(xp_y - yp_x\right) = -\boldsymbol{\omega}\cdot\mathbf{L}$$

where $\mathbf{L}$ is the angular momentum in the inertial frame. So the Hamiltonian in the rotating frame is

$$H' = H - \boldsymbol{\omega}\cdot\mathbf{L}.$$

As pointed out in *Lagrangian and Hamiltonian Mechanics*, H' is like a "free energy."

Exercise 6.06

The equations of motion for a particle of mass m and charge e moving in a uniform magnetic field B which points in the z-direction can be obtained from a Lagrangian (see Exercise 3.14)

$$L = \tfrac{1}{2}m(\dot{x}^2 + \dot{y}^2 + \dot{z}^2) + (eB/2c)(x\dot{y} - y\dot{x}).$$

(a) Find the momenta (p_x, p_y, p_z) conjugate to (x, y, z).
(b) Find the Hamiltonian, expressing your answer first in terms of $(x, y, z, \dot{x}, \dot{y}, \dot{z})$ and then in terms of (x, y, z, p_x, p_y, p_z).
(c) Evaluate the Poisson brackets
 (i) $[m\dot{x}, m\dot{y}]$
 (ii) $[m\dot{x}, H]$

Solution

(a) The momenta (p_x, p_y, p_z) conjugate to (x, y, z) are

$$p_x = \frac{\partial L}{\partial \dot{x}} = m\dot{x} - \frac{eB}{2c}y, \quad p_y = \frac{\partial L}{\partial \dot{y}} = m\dot{y} + \frac{eB}{2c}x, \quad \text{and} \quad p_z = \frac{\partial L}{\partial \dot{z}} = m\dot{z}.$$

(b) The Hamiltonian is

$$\begin{aligned} H &= p_x\dot{x} + p_y\dot{y} + p_z\dot{z} - L \\ &= \left(m\dot{x} - \frac{eB}{2c}y\right)\dot{x} + \left(m\dot{y} + \frac{eB}{2c}x\right)\dot{y} + m\dot{z}^2 - \frac{1}{2}m(\dot{x}^2 + \dot{y}^2 + \dot{z}^2) - \frac{eB}{2c}(x\dot{y} - y\dot{x}) \\ &= \frac{1}{2}m(\dot{x}^2 + \dot{y}^2 + \dot{z}^2) \end{aligned}$$

and equals "kinetic energy." Writing H in terms of the appropriate canonical variables (x, y, z, p_x, p_y, p_z), we have

$$H = \frac{1}{2m}\left(p_x + \frac{eB}{2c}y\right)^2 + \frac{1}{2m}\left(p_y - \frac{eB}{2c}x\right)^2 + \frac{1}{2m}p_z^2.$$

(c) The Poisson brackets of the components of the "kinematic momentum" with one another are

$$\begin{aligned} [m\dot{x}, m\dot{y}] &= \left[p_x + \frac{eB}{2c}y, p_y - \frac{eB}{2c}x\right] \\ &= [p_x, p_y] - \frac{eB}{2c}[p_x, x] + \frac{eB}{2c}[y, p_y] - \left(\frac{eB}{2c}\right)^2[y, x] \\ &= \frac{eB}{c}, \end{aligned}$$

together with

$$[m\dot{y}, m\dot{z}] = [m\dot{z}, m\dot{x}] = 0.$$

The Poisson brackets of the components of the "kinematic momentum" with the Hamiltonian are then

$$\begin{aligned} [m\dot{x}, H] &= \left[m\dot{x}, \frac{1}{2}m\left(\dot{x}^2 + \dot{y}^2 + \dot{z}^2\right)\right] = [m\dot{x}, m\dot{y}]\dot{y} = \frac{eB}{c}\dot{y}, \\ [m\dot{y}, H] &= \left[m\dot{y}, \frac{1}{2}m\left(\dot{x}^2 + \dot{y}^2 + \dot{z}^2\right)\right] = [m\dot{y}, m\dot{x}]\dot{x} = -\frac{eB}{c}\dot{x}, \\ [m\dot{z}, H] &= 0. \end{aligned}$$

These lead directly to the equations of motion for a charged particle in a uniform magnetic field, namely

$$\frac{d}{dt}(m\dot{x}) = [m\dot{x}, H] = \frac{eB}{c}\dot{y}, \quad \frac{d}{dt}(m\dot{y}) = [m\dot{y}, H] = -\frac{eB}{c}\dot{x}, \quad \text{and} \quad \frac{d}{dt}(m\dot{z}) = [m\dot{z}, H] = 0.$$

Exercise 6.07

Consider the one-dimensional motion of a particle in the following potential wells, in each case sketching representative trajectories in (x,p) phase space:

(a) an infinite square well $\quad V(x) = 0 \quad$ for $0 < x < a$
$\quad V(x) \to \infty \quad$ for $x \le 0$ and for $x > a$

(b) a bouncing ball $\quad V(x) = mgx \quad$ for $x > 0$
$\quad V(x) \to \infty \quad$ for $x \le 0$

(c) a simple harmonic oscillator $\quad V(x) = \frac{1}{2}kx^2$

(d) a double well $\quad V(x) = -\frac{1}{2}kx^2 + \frac{1}{4}k\frac{x^4}{a^2}$

Solution

(a) A particle of mass m and energy E in an "infinite square well potential" moves with constant momentum $\sqrt{2mE}$ from $x = 0$ to $x = a$ where it reflects, its momentum changing discontinuously to $-\sqrt{2mE}$. It then moves with constant momentum $-\sqrt{2mE}$ from $x = a$ back to $x = 0$ where it reflects, its momentum changing discontinuously back to $\sqrt{2mE}$, and the cycle is repeated. A representative phase space trajectory is shown in Fig. 1.

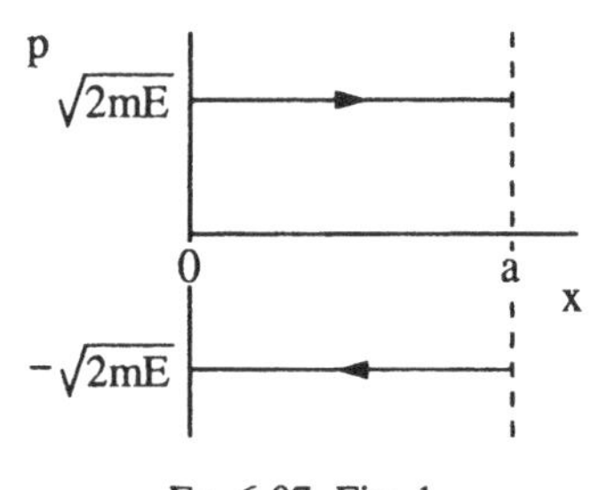

Ex. 6.07, Fig. 1

(b) A particle of mass m and energy E in a "bouncing ball potential" moves, for $x > 0$, so as to keep the total energy constant,

$$\frac{p^2}{2m} + mgx = E.$$

The trajectories in phase space are parabolas with x-intercept E/mg and p-intercepts $\pm\sqrt{2mE}$. At $x = 0$ the particle reflects, its momentum changing discontinuously from $-\sqrt{2mE}$ to $+\sqrt{2mE}$. A representative phase space trajectory is shown in Fig. 2.

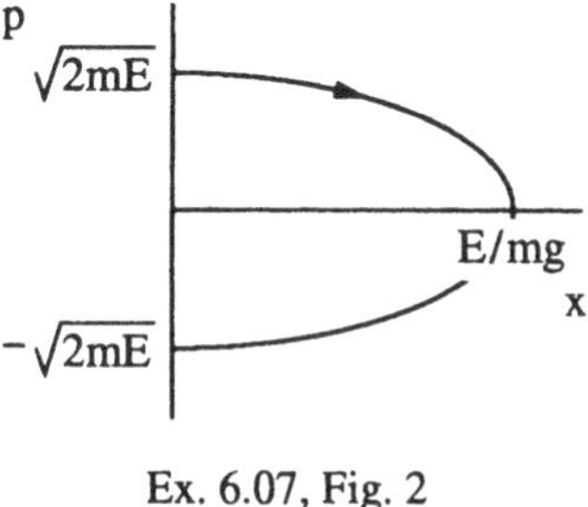

Ex. 6.07, Fig. 2

(c) A particle of mass m and energy E in a "simple harmonic oscillator potential" moves so as to keep the total energy constant,

$$\frac{p^2}{2m} + \frac{1}{2}kx^2 = E.$$

The trajectories in phase space are ellipses with x-intercepts $\pm\sqrt{2E/k}$ and p-intercepts $\pm\sqrt{2mE}$. A representative phase space trajectory is shown in Fig. 3.

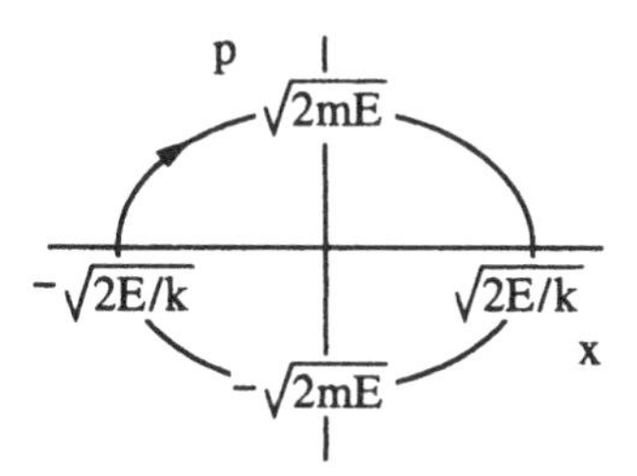

Ex. 6.07, Fig. 3

(d) A particle of mass m and energy E in a "double well potential" moves so as to keep the total energy constant,

$$H = \frac{p^2}{2m} - \frac{1}{2}kx^2 + \frac{1}{4}k\frac{x^4}{a^2} = E.$$

The equilibrium points satisfy

$$\frac{\partial H}{\partial p} = \frac{p}{m} = 0, \qquad \frac{\partial H}{\partial q} = -kx + k\frac{x^3}{a^2} = 0.$$

They are $(-a,0)$, $(0,0)$, and $(+a,0)$. The trajectories near $(0,0)$ are given by

$$\frac{p^2}{2m} - \frac{1}{2}kx^2 \approx E$$

and are hyperbolas; $(0,0)$ is a hyperbolic point. The trajectories near $(\mp a,0)$ are given by

$$\frac{p^2}{2m} + k(x \pm a)^2 \approx E + \frac{1}{4}ka^2$$

and are ellipses; $(\mp a,0)$ are elliptic points. Representative phase space trajectories are shown in Fig. 4. The trajectories for which $E = 0$ are the separatrices.

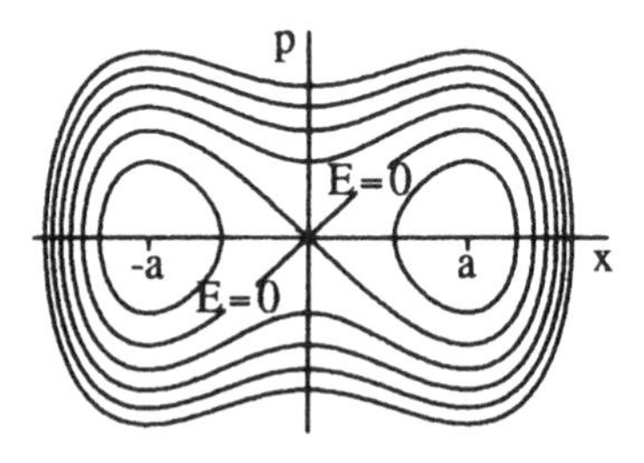

Ex. 6.07, Fig. 4

Exercise 6.08

In the potential wells of Exercise 6.07 the motion is periodic but not necessarily simple harmonic. The action variable I is defined by

$$I = \frac{1}{2\pi}\oint p\,dx$$

where p is the momentum, and the integration is over a single period of the motion (see Chapter IX, *Lagrangian and Hamiltonian Mechanics*, for further details).
(a) Show that the action variable is the *area enclosed* by the orbit in phase space divided by 2π and is given by

$$I = \frac{1}{\pi}\int_{x_1(E)}^{x_2(E)} \sqrt{2m(E - V(x))}\,dx,$$

where E is the total energy, $V(x)$ is the potential energy, and $x_1(E)$ and $x_2(E)$ are the lower and upper turning points of the motion.
(b) Show that the period τ of the motion is given by

$$\tau = 2\pi \frac{dI}{dE}.$$

Solution

(a) Consider a particle of mass m and energy E in a potential well $V(x)$. If $V(x) = E$ for $x = x_1$ and $x = x_2$, and $V(x) < E$ for $x_1 < x < x_2$, the particle oscillates back and forth between turning points x_1 and x_2. The momentum p at position x is

$$p = \sqrt{2m(E - V(x))}$$

with the square root positive as x increases from x_1 to x_2 and negative as x decreases from x_2 back to x_1. The action variable I(E) is

$$I = \frac{1}{2\pi}\oint p\,dx = \frac{1}{\pi}\int_{x_1(E)}^{x_2(E)} \sqrt{2m(E - V(x))}\,dx$$

with the x-increasing and x-decreasing halves of the cycle giving equal contributions.

(b) Consider the derivative of I with respect to E,

$$\begin{aligned}\frac{dI}{dE} &= \frac{1}{\pi}\int_{x_1(E)}^{x_2(E)} \frac{d}{dE}\sqrt{2m(E - V(x))}\,dx \\ &\quad + \frac{1}{\pi}\sqrt{2m(E - V(x_2))}\,\frac{dx_2}{dE} - \frac{1}{\pi}\sqrt{2m(E - V(x_1))}\,\frac{dx_1}{dE}.\end{aligned}$$

The second and third terms, which come from the dependence of the limits on E, give zero contribution since the turning points satisfy $\sqrt{2m(E - V(x_{1,2}))} = 0$. We are left with

$$\frac{dI}{dE} = \frac{1}{\pi}\int_{x_1(E)}^{x_2(E)} \sqrt{\frac{m}{2(E - V(x))}}\,dx = \frac{1}{\pi}\int_{x_1(E)}^{x_2(E)} \frac{dx}{v(x)}$$

where $v(x) = \sqrt{2(E - V(x))/m}$ is the speed of the particle at position x. We recognize the last integral as the time for the particle to go from x_1 to x_2, half a period τ, so we have

$$\tau = 2\pi \frac{dI}{dE}.$$

Exercise 6.09

(a) Evaluate the action variable $I(E)$ (see Exercise 6.08) for:

(i) an infinite square well $V(x) = 0$ for $0 < x < a$
$V(x) \to \infty$ for $x \leq 0$ and for $x > a$

(ii) a bouncing ball $V(x) = mgx$ for $x > 0$
$V(x) \to \infty$ for $x \leq 0$

(iii) a simple harmonic oscillator $V(x) = \frac{1}{2}kx^2$

and use your results to find the periods of the motions.

(b) In the "old quantum mechanics" of Bohr and Sommerfeld the action variable I was quantized in units of $\hbar$ (Planck's constant divided by 2π). What does this give for the energy levels of the systems of part (a)?

Solution

(a) (i) For a particle of mass m in an "infinite square well potential" the action variable is

$$I = \frac{1}{\pi}\int_0^a \sqrt{2mE}\,dx = \frac{a}{\pi}\sqrt{2mE},$$

and the period is

$$\tau = 2\pi\frac{dI}{dE} = a\sqrt{\frac{2m}{E}}.$$

To check, note that this equals $2a/v$, where $v = \sqrt{2E/m}$ is the constant speed of the particle, and is indeed the time required to make the round trip.

(ii) For a particle of mass m in a "bouncing ball potential" the action variable is

$$I = \frac{1}{\pi}\int_0^{E/mg} \sqrt{2m(E - mgx)}\,dx = \frac{2}{3\pi}\frac{\sqrt{2m}}{mg}E^{3/2},$$

and the period is

$$\tau = 2\pi\frac{dI}{dE} = 2\frac{\sqrt{2mE}}{mg}.$$

To check, note that in the half period "top to bottom" the particle falls a distance E/mg in a time $\tau/2$. These are related by the well-known equation for free fall,

$$\frac{E}{mg} = \frac{1}{2}g\left(\frac{\tau}{2}\right)^2,$$

which, on rearranging, gives the preceding expression for τ.

(iii) For a particle of mass m in a "harmonic oscillator potential" the action variable is

$$I = \frac{1}{\pi}\int_{-\sqrt{2E/k}}^{\sqrt{2E/k}} \sqrt{2m(E - \tfrac{1}{2}kx^2)}\, dx.$$

The integration can be performed by setting

$$x = \sqrt{2E/k}\sin\phi \quad \text{and} \quad dx = \sqrt{2E/k}\cos\phi\, d\phi$$

to give

$$I = \frac{2E}{\pi}\sqrt{\frac{m}{k}}\int_{-\pi/2}^{\pi/2}\cos^2\phi\, d\phi = \frac{E}{\omega}$$

where $\omega = \sqrt{k/m}$ is the angular frequency. The period is

$$\tau = 2\pi\frac{dI}{dE} = \frac{2\pi}{\omega}$$

which should be familiar.

(b) (i) For the infinite square well, Bohr-Sommerfeld quantization says

$$I_n = \frac{a}{\pi}\sqrt{2mE_n} = n\hbar \quad \text{where} \quad n = 0,1,2,\cdots$$

which gives the energy levels

$$E_n = n^2\frac{\hbar^2\pi^2}{2ma^2}.$$

Compare this with the result given by quantum mechanics, $E_n = (n+1)^2\dfrac{\hbar^2\pi^2}{2ma^2}$ where again $n = 0,1,2,\cdots$.

(ii) For the bouncing ball, Bohr-Sommerfeld quantization says

$$I_n = \frac{2}{3\pi}\frac{\sqrt{2m}}{mg}E_n^{3/2} = n\hbar \quad \text{where} \quad n = 0,1,2,\cdots$$

which gives the energy levels

$$E_n(BS) = \left(\frac{3\pi n}{2}\right)^{2/3}\left(\frac{mg^2\hbar^2}{2}\right)^{1/3}.$$

Compare this with the result given by the WKB approximation in quantum mechanics, $E_n(\text{WKB}) = \left(3\pi(n+\frac{3}{4})/2\right)^{2/3}\left(mg^2\hbar^2/2\right)^{1/3}$ where $n = 0,1,2,\cdots$. Also, compare it with the exact result given by quantum mechanics, $E_n(\text{exact}) = \lambda_n\left(mg^2\hbar^2/2\right)^{1/3}$ where the λ_n are the roots of the Airy function, $\text{Ai}(-\lambda_n) = 0$. See Table 1.

Table 1, Energy of quantum "bouncing ball," in units of $\left(mg^2\hbar^2/2\right)^{1/3}$

n=	$E_n(\text{BS}) =$	$E_n(\text{WKB}) =$	$E_n(\text{exact}) =$
0	0	2.3203	2.3381
1	2.8108	4.0818	4.0879
2	4.4618	5.5172	5.5206
3	5.8467	6.7845	6.7867
4	7.0827	7.9425	7.9441
⋮			
8	11.2431	11.9353	11.9360

(iii) For the simple harmonic oscillator, Bohr-Sommerfeld quantization says

$$I_n = \frac{E_n}{\omega} = n\hbar \quad \text{where} \quad n = 0,1,2,\cdots$$

which gives the energy levels

$$E_n = n\hbar\omega.$$

Compare this with the result given by quantum mechanics, $E = (n+\frac{1}{2})\hbar\omega$ where $n = 0,1,2,\cdots$.

Exercise 6.10

(a) Use the definition of the Poisson bracket to establish Poisson bracket properties 1, 2, 3, and also the Jacobi identity 4.
(b) Show that these four properties also hold for the commutator bracket.

Solution

(a) For Poisson brackets we have

$$\begin{aligned}\left[u_1+u_2,v\right] &= \sum_{a=1}^{f}\left(\frac{\partial(u_1+u_2)}{\partial q_a}\frac{\partial v}{\partial p_a}-\frac{\partial(u_1+u_2)}{\partial p_a}\frac{\partial v}{\partial q_a}\right)\\ &= \sum_{a=1}^{f}\left(\left(\frac{\partial u_1}{\partial q_a}+\frac{\partial u_2}{\partial q_a}\right)\frac{\partial v}{\partial p_a}-\left(\frac{\partial u_1}{\partial p_a}+\frac{\partial u_2}{\partial p_a}\right)\frac{\partial v}{\partial q_a}\right)\\ &= \sum_{a=1}^{f}\left(\frac{\partial u_1}{\partial q_a}\frac{\partial v}{\partial p_a}-\frac{\partial u_1}{\partial p_a}\frac{\partial v}{\partial q_a}\right)+\sum_{a=1}^{f}\left(\frac{\partial u_2}{\partial q_a}\frac{\partial v}{\partial p_a}-\frac{\partial u_2}{\partial p_a}\frac{\partial v}{\partial q_a}\right)\\ &= \left[u_1,v\right]+\left[u_2,v\right],\end{aligned}$$

$$\begin{aligned}\left[cu,v\right] &= \sum_{a=1}^{f}\left(\frac{\partial(cu)}{\partial q_a}\frac{\partial v}{\partial p_a}-\frac{\partial(cu)}{\partial p_a}\frac{\partial v}{\partial q_a}\right)\\ &= \sum_{a=1}^{f}\left(c\frac{\partial u}{\partial q_a}\frac{\partial v}{\partial p_a}-c\frac{\partial u}{\partial p_a}\frac{\partial v}{\partial q_a}\right)\\ &= c\sum_{a=1}^{f}\left(\frac{\partial u}{\partial q_a}\frac{\partial v}{\partial p_a}-\frac{\partial u}{\partial p_a}\frac{\partial v}{\partial q_a}\right)\\ &= c\left[u,v\right],\end{aligned}$$

$$\begin{aligned}\left[u,v\right] &= \sum_{a=1}^{f}\left(\frac{\partial u}{\partial q_a}\frac{\partial v}{\partial p_a}-\frac{\partial u}{\partial p_a}\frac{\partial v}{\partial q_a}\right)\\ &= -\sum_{a=1}^{f}\left(\frac{\partial v}{\partial q_a}\frac{\partial u}{\partial p_a}-\frac{\partial v}{\partial p_a}\frac{\partial u}{\partial q_a}\right)\\ &= -\left[v,u\right],\end{aligned}$$

$$\begin{aligned}\left[u,vw\right] &= \sum_{a=1}^{f}\left(\frac{\partial u}{\partial q_a}\frac{\partial(vw)}{\partial p_a}-\frac{\partial u}{\partial p_a}\frac{\partial(vw)}{\partial q_a}\right)\\ &= \sum_{a=1}^{f}\left(\frac{\partial u}{\partial q_a}\left(\frac{\partial v}{\partial p_a}w+v\frac{\partial w}{\partial p_a}\right)-\frac{\partial u}{\partial p_a}\left(\frac{\partial v}{\partial q_a}w+v\frac{\partial w}{\partial q_a}\right)\right)\\ &= \sum_{a=1}^{f}\left(\frac{\partial u}{\partial q_a}\frac{\partial v}{\partial p_a}-\frac{\partial u}{\partial p_a}\frac{\partial v}{\partial q_a}\right)w+v\sum_{a=1}^{f}\left(\frac{\partial u}{\partial q_a}\frac{\partial w}{\partial p_a}-\frac{\partial u}{\partial p_a}\frac{\partial w}{\partial q_a}\right)\\ &= \left[u,v\right]w+v\left[u,w\right].\end{aligned}$$

This proves the first three properties of Poisson brackets. To prove the fourth property, the Jacobi identity, we start with

$$\left[v,w\right]=\sum_{a=1}^{f}\left(\frac{\partial v}{\partial q_a}\frac{\partial w}{\partial p_a}-\frac{\partial v}{\partial p_a}\frac{\partial w}{\partial q_a}\right)$$

and work out

$$[u,[v,w]]=\sum_{b=1}^{f}\sum_{a=1}^{f}\left(\frac{\partial u}{\partial q_b}\frac{\partial}{\partial p_b}\left(\frac{\partial v}{\partial q_a}\frac{\partial w}{\partial p_a}-\frac{\partial v}{\partial p_a}\frac{\partial w}{\partial q_a}\right)-\frac{\partial u}{\partial p_b}\frac{\partial}{\partial q_b}\left(\frac{\partial v}{\partial q_a}\frac{\partial w}{\partial p_a}-\frac{\partial v}{\partial p_a}\frac{\partial w}{\partial q_a}\right)\right)$$

$$=\sum_{b=1}^{f}\sum_{a=1}^{f}\left(\frac{\partial u}{\partial q_b}\left(\frac{\partial^2 v}{\partial p_b\partial q_a}\frac{\partial w}{\partial p_a}+\frac{\partial v}{\partial q_a}\frac{\partial^2 w}{\partial p_b\partial p_a}-\frac{\partial^2 v}{\partial p_b\partial p_a}\frac{\partial w}{\partial q_a}-\frac{\partial v}{\partial p_a}\frac{\partial^2 w}{\partial p_b\partial q_a}\right)\right.$$

$$\left.-\frac{\partial u}{\partial p_b}\left(\frac{\partial^2 v}{\partial q_b\partial q_a}\frac{\partial w}{\partial p_a}+\frac{\partial v}{\partial q_a}\frac{\partial^2 w}{\partial q_b\partial p_a}-\frac{\partial^2 v}{\partial q_b\partial p_a}\frac{\partial w}{\partial q_a}-\frac{\partial v}{\partial p_a}\frac{\partial^2 w}{\partial q_b\partial q_a}\right)\right).$$

Permuting u, v, and w cyclically, we obtain similar expressions for $[v,[w,u]]$ and for $[w,[u,v]]$. Adding these , we find that all terms on the right-hand side cancel and we thus obtain the Jacobi identity

$$[u,[v,w]]+[v,[w,u]]+[w,[u,v]]=0.$$

(b For commutator brackets we have

$$\begin{aligned}[u_1+u_2,v]&=(u_1+u_2)v-v(u_1+u_2)\\&=(u_1v-vu_1)+(u_2v-vu_2)\\&=[u_1,v]+[u_2,v],\end{aligned}$$

$$\begin{aligned}[cu,v]&=(cu)v-v(cu)\\&=c(uv-vu)\\&=c[u,v],\end{aligned}$$

$$\begin{aligned}[u,v]&=uv-vu\\&=-(vu-uv)\\&=-[v,u],\end{aligned}$$

$$\begin{aligned}[u,vw]&=u(vw)-(vw)u\\&=(uv-vu)w+v(uw-wu)\\&=[u,v]w+v[u,w].\end{aligned}$$

To prove the Jacobi identity for commutators, note that

$$\begin{aligned}[u,[v,w]]&=u(vw-wv)-(vw-wv)u,\\ [v,[w,u]]&=v(wu-uw)-(wu-uw)v,\\ [w,[u,v]]&=w(uv-vu)-(uv-vu)w.\end{aligned}$$

Adding these, we find that all terms on the right cancel and we obtain the Jacobi identity.

Exercise 6.11

Let $f(q(t), p(t))$ be some function of the canonical variables, and f_0 its value at time 0.
(a) Show, if the Hamiltonian H is time-independent, that the function f at time t is given by

$$f = f_0 + t[f_0, H] + (t^2/2!)[[f_0, H], H] + (t^3/3!)[[[f_0, H], H], H] + \cdots.$$

(b) A particle of mass m moving in one dimension x is acted on by a constant force F. The Hamiltonian is $H = p^2/2m - Fx$. Suppose that at time 0 the particle is at x_0 with momentum p_0. Use the result of (a) to find the position x and momentum p at time t.
(c) A particle of mass m moving in one dimension x is in a simple harmonic oscillator well. The Hamiltonian is $H = p^2/2m + m\omega^2 x^2/2$. Suppose that at time 0 the particle is at x_0 with momentum p_0. Use the result of (a) to find the position x and momentum p at time t.

Solution

(a) We expand f in a Taylor series,

$$f = f_0 + t\left(\frac{df}{dt}\right)_0 + \frac{t^2}{2!}\left(\frac{d^2f}{dt^2}\right)_0 + \cdots.$$

The successive time derivatives of f are given by

$$\frac{df}{dt} = [f, H], \quad \frac{d^2f}{dt^2} = \left[\frac{df}{dt}, H\right] = [[f, H], H], \quad \cdots$$

so we have

$$f = f_0 + t[f_0, H] + \frac{t^2}{2!}[[f_0, H], H] + \cdots$$

as required.

(b) Consider a particle of mass m moving in one dimension and acted on by a constant force F. The Hamiltonian is

$$H = \frac{p^2}{2m} - Fx.$$

Taking $f = x$ we have

$$[x,H] = \left[x, \frac{p^2}{2m} - Fx\right] = \frac{1}{2m}\left[x, p^2\right] = \frac{p}{m}[x,p] = \frac{p}{m},$$

$$[[x,H],H] = \left[\frac{p}{m}, \frac{p^2}{2m} - Fx\right] = -\frac{F}{m}[p,x] = \frac{F}{m}.$$

Since this last Poisson bracket is a constant, the higher order Poisson brackets give zero and we have

$$x = x_0 + \frac{p_0}{m}t + \frac{1}{2}\frac{F}{m}t^2.$$

Taking $f = p$ we have

$$[p,H] = \left[p, \frac{p^2}{2m} - Fx\right] = -F[p,x] = F.$$

Since this Poisson bracket is a constant, the higher order Poisson brackets give zero and we have

$$p = p_0 + Ft.$$

These are the well-known expressions for $x(t)$ and $p(t)$ for motion at constant acceleration F/m.

(c) Consider a particle of mass m moving in one dimension in a simple harmonic oscillator well. The Hamiltonian is

$$H = \frac{p^2}{2m} + \frac{1}{2}m\omega^2x^2.$$

Taking $f = x$ we have

$$[x,H] = \left[x, \frac{p^2}{2m} + \frac{1}{2}m\omega^2x^2\right] = \frac{1}{2m}\left[x, p^2\right] = \frac{p}{m}[x,p] = \frac{p}{m},$$

$$[[x,H],H] = \left[\frac{p}{m}, \frac{p^2}{2m} + \frac{1}{2}m\omega^2x^2\right] = \frac{1}{2}\omega^2\left[p, x^2\right] = \omega^2x[p,x] = -\omega^2x,$$

$$[[[x,H],H],H] = -\omega^2[x,H] = -\omega^2\frac{p}{m},$$

and so forth. We thus find

$$x = x_0\left(1 - \frac{1}{2!}\omega^2t^2 + \cdots\right) + \frac{p_0}{m\omega}\left(\omega t - \frac{1}{3!}\omega^3t^3 + \cdots\right) = x_0\cos\omega t + \frac{p_0}{m\omega}\sin\omega t.$$

Taking $f = p$ we have

$$[p,H] = \left[p, \frac{p^2}{2m} + \frac{1}{2}m\omega^2x^2\right] = \frac{1}{2}m\omega^2[p,x^2] = m\omega^2x[p,x] = -m\omega^2x,$$
$$[[p,H],H] = -m\omega^2[x,H] = -\omega^2p,$$
$$[[[p,H],H],H] = -\omega^2[p,H] = m\omega^4x,$$

and so forth. We thus find

$$p = -m\omega x_0\left(\omega t - \frac{1}{3!}\omega^3t^3 + \cdots\right) + p_0\left(1 - \frac{1}{2!}\omega^2t^2 + \cdots\right) = -m\omega x_0 \sin\omega t + p_0 \cos\omega t.$$

These are the well-known expressions for $x(t)$ and $p(t)$ for simple harmonic motion.

Exercise 6.12

The Hamiltonian for a simple harmonic oscillator is

$$H = \frac{p^2}{2m} + \frac{1}{2}m\omega^2x^2.$$

Introduce the complex quantities

$$a = \sqrt{\frac{m\omega}{2}}\left(x + \frac{ip}{m\omega}\right) \quad \text{and} \quad a^* = \sqrt{\frac{m\omega}{2}}\left(x - \frac{ip}{m\omega}\right).$$

(a) Express H in terms of a and a^*.
(b) Evaluate the Poisson brackets $[a,a^*]$, $[a,H]$, and $[a^*,H]$.
(c) Write down and solve the equations of motion for a and a^*.

Solution

(a) We have

$$a^*a = \frac{m\omega}{2}\left(x^2 + \frac{p^2}{m^2\omega^2}\right),$$

so the Hamiltonian for a simple harmonic oscillator can be written

$$H = \frac{p^2}{2m} + \frac{1}{2}m\omega^2x^2 = \omega a^*a.$$

(b) The required Poisson brackets are

$$\left[a,a^*\right] = \frac{m\omega}{2}\left[x + \frac{ip}{m\omega}, x - \frac{ip}{m\omega}\right] = -i,$$
$$[a,H] = \omega\left[a,a^*a\right] = \omega\left[a,a^*\right]a = -i\omega a,$$
$$\left[a^*,H\right] = \omega\left[a^*,a^*a\right] = \omega a^*\left[a^*,a\right] = i\omega a^*.$$

(c) The time rates of change of a and of a^* are thus

$$\frac{da}{dt} = [a,H] = -i\omega a \quad \text{and} \quad \frac{da^*}{dt} = \left[a^*,H\right] = i\omega a^*.$$

These can be integrated to give

$$a = a_0 e^{-i\omega t} \quad \text{and} \quad a^* = a_0^* e^{i\omega t}$$

where a_0 and a_0^* are the initial values of a and a^*.

This provides yet another way to obtain the general solution to the harmonic oscillator problem,

$$x = \frac{1}{\sqrt{2m\omega}}\left(a + a^*\right) = \frac{1}{\sqrt{2m\omega}}\left(a_0 e^{-i\omega t} + a_0^* e^{i\omega t}\right),$$
$$p = -i\sqrt{\frac{m\omega}{2}}\left(a - a^*\right) = -i\sqrt{\frac{m\omega}{2}}\left(a_0 e^{-i\omega t} - a_0^* e^{i\omega t}\right).$$

Exercise 6.13

(a) Evaluate the set of Poisson brackets for a component of the radius vector $\mathbf{r} = (x,y,z)$ with a component of the angular momentum $\mathbf{L} = (L_x, L_y, L_z)$. Also evaluate those for a component of the linear momentum $\mathbf{p} = (p_x, p_y, p_z)$ with a component of the angular momentum. Show that the results can be put in the form

$$[\mathbf{r}, \mathbf{L}\cdot\mathbf{n}] = \mathbf{n}\times\mathbf{r} \qquad [\mathbf{p}, \mathbf{L}\cdot\mathbf{n}] = \mathbf{n}\times\mathbf{p}$$

where **n** is an arbitrary constant vector.

(b) Use the results of (a) to show that the Poisson bracket of a component of the angular momentum with an arbitrary scalar function of **r** and **p**, of the form $a(r^2, \mathbf{r}\cdot\mathbf{p}, p^2)$, is zero.

(c) Use the results of (a) to show that the Poisson bracket of a component of the angular momentum with an arbitrary vector function of **r** and **p**, of the form

$$\mathbf{A} = a_1\mathbf{r} + a_2\mathbf{p} + a_3\mathbf{r}\times\mathbf{p},$$

is given by

$$[\mathbf{A},\mathbf{L}\cdot\mathbf{n}] = \mathbf{n}\times\mathbf{A}.$$

(d) Show that the Poisson bracket of the square of the angular momentum $L^2 = L_x^2 + L_y^2 + L_z^2$ with an arbitrary vector function $\mathbf{A}$ of $\mathbf{r}$ and $\mathbf{p}$ is given by

$$[\mathbf{A},L^2] = 2\mathbf{L}\times\mathbf{A}.$$

Solution

(a) We have

$$\begin{aligned}
&[x,L_x] = [x,yp_z - zp_y] = 0, \quad [y,L_x] = [y,yp_z - zp_y] = -z, \quad [z,L_x] = [z,yp_z - zp_y] = y,\\
&[x,L_y] = [x,zp_x - xp_z] = z, \quad [y,L_y] = [y,zp_x - xp_z] = 0, \quad [z,L_y] = [z,zp_x - xp_z] = -x,\\
&[x,L_z] = [x,xp_y - yp_x] = -y, \quad [y,L_z] = [y,xp_y - yp_x] = x, \quad [z,L_z] = [z,xp_y - yp_x] = 0.
\end{aligned}$$

Now let $\mathbf{n} = n_x\mathbf{i} + n_y\mathbf{j} + n_z\mathbf{k}$ be an arbitrary constant vector. The preceding equations can then be written

$$\begin{aligned}
[x,\mathbf{L}\cdot\mathbf{n}] &= n_y z - n_z y = (\mathbf{n}\times\mathbf{r})_x,\\
[y,\mathbf{L}\cdot\mathbf{n}] &= n_z x - n_x z = (\mathbf{n}\times\mathbf{r})_y,\\
[z,\mathbf{L}\cdot\mathbf{n}] &= n_x y - n_y x = (\mathbf{n}\times\mathbf{r})_z.
\end{aligned}$$

These, in turn, can be summarized as

$$[\mathbf{r},\mathbf{L}\cdot\mathbf{n}] = \mathbf{n}\times\mathbf{r}.$$

The set of Poisson brackets for a component of $\mathbf{p}$ with a component of $\mathbf{L}$ can be found similarly. It is, however, simpler to obtain these by interchanging $\mathbf{r}$ and $\mathbf{p}$ in the above equations. There are two compensating sign changes, one in the definition of the Poisson bracket and one in the definition of the angular momentum. We thus find

$$[\mathbf{p},\mathbf{L}\cdot\mathbf{n}] = \mathbf{n}\times\mathbf{p}.$$

(b) To evaluate the Poisson bracket of an arbitrary function of r^2, p^2, and $\mathbf{r}\cdot\mathbf{p}$ with $\mathbf{L}\cdot\mathbf{n}$, we first consider the Poisson brackets of r^2, p^2, and $\mathbf{r}\cdot\mathbf{p}$ with $\mathbf{L}\cdot\mathbf{n}$. We have

$$\begin{aligned}
[r^2,\mathbf{L}\cdot\mathbf{n}] &= [\mathbf{r}\cdot\mathbf{r},\mathbf{L}\cdot\mathbf{n}] = 2\mathbf{r}\cdot[\mathbf{r},\mathbf{L}\cdot\mathbf{n}] = 2\mathbf{r}\cdot\mathbf{n}\times\mathbf{r} = 0,\\
[p^2,\mathbf{L}\cdot\mathbf{n}] &= [\mathbf{p}\cdot\mathbf{p},\mathbf{L}\cdot\mathbf{n}] = 2\mathbf{p}\cdot[\mathbf{p},\mathbf{L}\cdot\mathbf{n}] = 2\mathbf{p}\cdot\mathbf{n}\times\mathbf{p} = 0,\\
[\mathbf{r}\cdot\mathbf{p},\mathbf{L}\cdot\mathbf{n}] &= \mathbf{r}\cdot[\mathbf{p},\mathbf{L}\cdot\mathbf{n}] + \mathbf{p}\cdot[\mathbf{r},\mathbf{L}\cdot\mathbf{n}] = \mathbf{r}\cdot\mathbf{n}\times\mathbf{p} + \mathbf{p}\cdot\mathbf{n}\times\mathbf{r} = 0.
\end{aligned}$$

We then need the Poisson bracket $[u,v]$ where u is some function $u(\xi_1,\xi_2,\cdots)$ of functions $\xi_1,\xi_2,\cdots$ of the canonical variables. For this we have

$$[u,v] = \sum_{a=1}^{f}\left(\frac{\partial u}{\partial q_a}\frac{\partial v}{\partial p_a} - \frac{\partial u}{\partial p_a}\frac{\partial v}{\partial q_a}\right) = \sum_{a=1}^{f}\sum_{i=1}^{n}\left(\frac{\partial u}{\partial \xi_i}\frac{\partial \xi_i}{\partial q_a}\frac{\partial v}{\partial p_a} - \frac{\partial u}{\partial \xi_i}\frac{\partial \xi_i}{\partial p_a}\frac{\partial v}{\partial q_a}\right) = \sum_{i=1}^{n}\frac{\partial u}{\partial \xi_i}[\xi_i, v],$$

the second equality following from the chain rule for partial derivatives. Thus

$$[a(r^2, \mathbf{r}\cdot\mathbf{p}, p^2), \mathbf{L}\cdot\mathbf{n}] = \frac{\partial a}{\partial r^2}[r^2, \mathbf{L}\cdot\mathbf{n}] + \frac{\partial a}{\partial(\mathbf{r}\cdot\mathbf{p})}[\mathbf{r}\cdot\mathbf{p}, \mathbf{L}\cdot\mathbf{n}] + \frac{\partial a}{\partial p^2}[p^2, \mathbf{L}\cdot\mathbf{n}] = 0.$$

(c) We have

$$\begin{aligned}
[\mathbf{A}, \mathbf{L}\cdot\mathbf{n}] &= [a_1\mathbf{r} + a_2\mathbf{p} + a_3\mathbf{r}\times\mathbf{p}, \mathbf{L}\cdot\mathbf{n}] \\
&= a_1[\mathbf{r}, \mathbf{L}\cdot\mathbf{n}] + a_2[\mathbf{p}, \mathbf{L}\cdot\mathbf{n}] + a_3[\mathbf{r}\times\mathbf{p}, \mathbf{L}\cdot\mathbf{n}] \\
&= a_1\mathbf{n}\times\mathbf{r} + a_2\mathbf{n}\times\mathbf{p} + a_3\mathbf{r}\times[\mathbf{p}, \mathbf{L}\cdot\mathbf{n}] + a_3[\mathbf{r}, \mathbf{L}\cdot\mathbf{n}]\times\mathbf{p} \\
&= a_1\mathbf{n}\times\mathbf{r} + a_2\mathbf{n}\times\mathbf{p} + a_3(\mathbf{r}\times(\mathbf{n}\times\mathbf{p}) + (\mathbf{n}\times\mathbf{r})\times\mathbf{p}) \\
&= a_1\mathbf{n}\times\mathbf{r} + a_2\mathbf{n}\times\mathbf{p} + a_3\mathbf{n}\times(\mathbf{r}\times\mathbf{p}) \\
&= \mathbf{n}\times\mathbf{A}\,.
\end{aligned}$$

This result can be obtained much more easily by looking ahead to some results in Chapter VII, *Lagrangian and Hamiltonian Mechanics*. The change $\delta\mathbf{A}$ in a vector $\mathbf{A}$ under an infinitesimal rotation $\delta\boldsymbol{\theta}$ is

$$\delta\mathbf{A} = \delta\boldsymbol{\theta}\times\mathbf{A}.$$

On the other hand, this is a canonical transformation generated by $\mathbf{L}\cdot\delta\boldsymbol{\theta}$, so

$$\delta\mathbf{A} = [\mathbf{A}, \mathbf{L}\cdot\delta\boldsymbol{\theta}].$$

Comparing these expressions, we find

$$[\mathbf{A}, \mathbf{L}\cdot\delta\boldsymbol{\theta}] = \delta\boldsymbol{\theta}\times\mathbf{A}$$

as required.

(d) We have

$$\begin{aligned}
[\mathbf{A}, L^2] &= [\mathbf{A}, L_x^2 + L_y^2 + L_z^2] \\
&= 2L_x[\mathbf{A}, L_x] + 2L_y[\mathbf{A}, L_y] + 2L_z[\mathbf{A}, L_z] \\
&= 2L_x(\mathbf{i}\times\mathbf{A}) + 2L_y(\mathbf{j}\times\mathbf{A}) + 2L_z(\mathbf{k}\times\mathbf{A}) \\
&= 2\mathbf{L}\times\mathbf{A}\,.
\end{aligned}$$

Exercise 6.14

Consider motion of a particle of mass m in an isotropic harmonic oscillator potential $V = \frac{1}{2}kr^2$ and take the orbital plane to be the x-y plane. The Hamiltonian is then

$$H = S_0 = \frac{1}{2m}(p_x^2 + p_y^2) + \frac{1}{2}k(x^2 + y^2).$$

Introduce the three quantities

$$S_1 = \frac{1}{2m}(p_x^2 - p_y^2) + \frac{1}{2}k(x^2 - y^2), \quad S_2 = \frac{1}{m}p_x p_y + kxy, \quad S_3 = \omega(xp_y - yp_x),$$

with $\omega = \sqrt{k/m}$.

(a) Show that

$$[S_0, S_i] = 0 \qquad i = 1,2,3$$

so (S_1, S_2, S_3) are constants of the motion.

(b) Show that

$$[S_1, S_2] = 2\omega S_3, \qquad [S_2, S_3] = 2\omega S_1, \qquad [S_3, S_1] = 2\omega S_2,$$

so $(2\omega)^{-1}(S_1, S_2, S_3)$ have the same Poisson bracket relations as the components of a "three-dimensional angular momentum."

(c) Show that

$$S_0^2 = S_1^2 + S_2^2 + S_3^2.$$

(The corresponding quantum relation has $(\hbar\omega)^2$ added to the right-hand side.)

Solution

(a) We have

$$\begin{aligned}
\left[S_0, S_1\right] &= \left[\frac{1}{2m}(p_x^2 + p_y^2) + \frac{1}{2}k(x^2 + y^2), \frac{1}{2m}(p_x^2 - p_y^2) + \frac{1}{2}k(x^2 - y^2)\right] \\
&= \frac{k}{4m}\left(\left[p_x^2, x^2\right] - \left[p_y^2, y^2\right] + \left[x^2, p_x^2\right] - \left[y^2, p_y^2\right]\right) \\
&= 0,
\end{aligned}$$

$$\begin{aligned}
\left[S_0,S_2\right] &= \left[\frac{1}{2m}(p_x^2+p_y^2)+\frac{1}{2}k(x^2+y^2),\frac{1}{m}p_xp_y+kxy\right] \\
&= \frac{k}{2m}\left(\left[p_x^2,xy\right]+\left[p_y^2,xy\right]+\left[x^2,p_xp_y\right]+\left[y^2,p_xp_y\right]\right) \\
&= \frac{k}{2m}\left(2yp_x\left[p_x,x\right]+2xp_y\left[p_y,y\right]+2xp_y\left[x,p_x\right]+2yp_x\left[y,p_y\right]\right) \\
&= \frac{k}{2m}\left(-2yp_x-2xp_y+2xp_y+2yp_x\right) \\
&= 0\,,
\end{aligned}$$

$$\begin{aligned}
\left[S_0,S_3\right] &= \left[\frac{1}{2m}(p_x^2+p_y^2)+\frac{1}{2}k(x^2+y^2),\omega(xp_y-yp_x)\right] \\
&= \frac{\omega}{2m}\left(\left[p_x^2,xp_y\right]-\left[p_y^2,yp_x\right]\right)+\frac{1}{2}k\omega\left(-\left[x^2,yp_x\right]+\left[y^2,xp_y\right]\right) \\
&= \frac{\omega}{2m}\left(2p_xp_y\left[p_x,x\right]-2p_xp_y\left[p_y,y\right]\right)+\frac{1}{2}k\omega\left(-2xy\left[x,p_x\right]+2xy\left[y,p_y\right]\right) \\
&= \frac{\omega}{2m}\left(-2p_xp_y+2p_xp_y\right)+\frac{1}{2}k\omega\left(-2xy+2xy\right) \\
&= 0\,.
\end{aligned}$$

So $\left(S_1,S_2,S_3\right)$ are constants of the motion.

(b) We have

$$\begin{aligned}
\left[S_1,S_2\right] &= \left[\frac{1}{2m}(p_x^2-p_y^2)+\frac{1}{2}(x^2-y^2),\frac{1}{m}p_xp_y+kxy\right] \\
&= \frac{k}{2m}\left(\left[p_x^2,xy\right]-\left[p_y^2,xy\right]+\left[x^2,p_xp_y\right]-\left[y^2,p_xp_y\right]\right) \\
&= \frac{k}{2m}\left(2yp_x\left[p_x,x\right]-2xp_y\left[p_y,y\right]+2xp_y\left[x,p_x\right]-2yp_x\left[y,p_y\right]\right) \\
&= 2\frac{k}{m}\left(-yp_x+xp_y\right) \\
&= 2\omega S_3\,,
\end{aligned}$$

$$\begin{aligned}
\left[S_2,S_3\right] &= \left[\frac{1}{m}p_xp_y+kxy,\omega(xp_y-yp_x)\right] \\
&= \frac{\omega}{m}\left(\left[p_xp_y,xp_y\right]-\left[p_xp_y,yp_x\right]\right)+k\omega\left(\left[xy,xp_y\right]-\left[xy,yp_x\right]\right) \\
&= \frac{\omega}{m}\left(p_y^2\left[p_x,x\right]-p_x^2\left[p_y,y\right]\right)+k\omega\left(x^2\left[y,p_y\right]-y^2\left[x,p_x\right]\right) \\
&= \frac{\omega}{m}\left(p_x^2-p_y^2\right)+k\omega\left(x^2-y^2\right) \\
&= 2\omega S_1\,,
\end{aligned}$$

$$\begin{aligned}
[S_3,S_1] &= \left[\omega(xp_y - yp_x), \frac{1}{2m}(p_x^2 - p_y^2) + \frac{1}{2}(x^2 - y^2)\right] \\
&= \frac{\omega}{2m}\left(\left[xp_y, p_x^2\right] + \left[yp_x, p_y^2\right]\right) - \frac{1}{2}k\omega\left(\left[xp_y, y^2\right] + \left[yp_x, x^2\right]\right) \\
&= \frac{\omega}{2m}\left(2p_xp_y[x,p_x] + 2p_xp_y[y,p_y]\right) - \frac{1}{2}k\omega\left(2xy[p_y,y] + 2xy[p_x,x]\right) \\
&= 2\frac{\omega}{m}p_xp_y + 2k\omega xy \\
&= 2\omega S_2 .
\end{aligned}$$

So $(2\omega)^{-1}(S_1,S_2,S_3)$ have the same Poisson bracket relations as the components of an abstract "three-dimensional angular momentum."

(c) Rather than approaching this part of the exercise head on, it is better to begin by observing that

$$\begin{aligned}
S_0 &= \left(\frac{1}{2m}p_x^2 + \frac{1}{2}kx^2\right) + \left(\frac{1}{2m}p_y^2 + \frac{1}{2}ky^2\right), \\
S_1 &= \left(\frac{1}{2m}p_x^2 + \frac{1}{2}kx^2\right) - \left(\frac{1}{2m}p_y^2 + \frac{1}{2}ky^2\right),
\end{aligned}$$

and thus

$$S_0^2 - S_1^2 = 4\left(\frac{1}{2m}p_x^2 + \frac{1}{2}kx^2\right)\left(\frac{1}{2m}p_y^2 + \frac{1}{2}ky^2\right).$$

Also, note that

$$\begin{aligned}
S_2^2 + S_3^2 &= \left(\frac{1}{m}p_xp_y + kxy\right)^2 + \frac{k}{m}\left(xp_y - yp_x\right)^2 \\
&= \frac{1}{m^2}p_x^2p_y^2 + k^2x^2y^2 + \frac{k}{m}\left(x^2p_y^2 + y^2p_x^2\right) \\
&= \left(\frac{1}{m}p_x^2 + kx^2\right)\left(\frac{1}{m}p_y^2 + ky^2\right).
\end{aligned}$$

Comparing, we see that

$$S_0^2 - S_1^2 = S_2^2 + S_3^2,$$

so

$$S_0^2 = S_1^2 + S_2^2 + S_3^2$$

as required.

In quantum mechanics the set of three quantities $(2\omega)^{-1}(S_1,S_2,S_3)$ have the same Poisson bracket relations with one another as in classical mechanics, so they are again like the components of "three-dimensional angular momentum." The square of the length of this "three-dimensional angular momentum" in quantum mechanics is

$$\left(\frac{S_1}{2\omega}\right)^2+\left(\frac{S_2}{2\omega}\right)^2+\left(\frac{S_3}{2\omega}\right)^2=\left(\frac{E}{2\omega}\right)^2-\frac{\hbar^2}{4}$$

where E is the energy of a two-dimensional harmonic oscillator with angular frequency ω (note the "extra" term $-\hbar^2/4$). Angular momentum theory shows that the eigenvalues of this quantity are $\lambda(\lambda+1)\hbar^2$ where $\lambda=0,\frac{1}{2},1,\frac{3}{2},\cdots$, so the eigenvalues of the energy of a two-dimensional harmonic oscillator are

$$E_n=\hbar\omega(2\lambda+1)=\hbar\omega(n+1) \quad \text{where} \quad n=0,1,2,\cdots.$$

Exercise 6.15

Consider motion of a particle of mass m in a gravitational potential $V=-k/r$ and take the orbital plane to be the x-y plane. The Hamiltonian is then

$$H=\frac{1}{2m}(p_x^2+p_y^2)-\frac{k}{r}$$

where now $r=\sqrt{x^2+y^2}$. The angular momentum vector points in the z-direction and has (z-) component

$$L=xp_y-yp_x,$$

and the Laplace-Runge-Lenz vector (see Exercise 1.12) lies in the x-y plane and has components

$$K_x=p_yL-mk\,x/r, \quad K_y=-p_xL-mk\,y/r.$$

(a) Show that

$$[L,H]=0, \quad [K_x,H]=0, \quad [K_y,H]=0,$$

so L, K_x, and K_y are constants of the motion.

(b) Show that

$$[K_x,L]=-K_y, \quad [K_y,L]=K_x, \quad [K_x,K_y]=-(2mH)L$$

(see Exercise 6.13 for some useful Poisson brackets). Now restrict yourself to bound states of energy $H = -|E|$ and show that $\left(\frac{K_x}{\sqrt{2m|E|}}, \frac{K_y}{\sqrt{2m|E|}}, L\right)$ have the same Poisson bracket relations as the components of a "three-dimensional angular momentum."

(c) Show that the square of the length of this "three-dimensional angular momentum" is $\frac{mk^2}{2|E|}$. (The corresponding quantum relation is $\frac{mk^2}{2|E|} - \frac{1}{4}\hbar^2$.)

Solution

(a) Since L is the z-component of the angular momentum and H is a scalar, the Poisson bracket of L with H is zero (see Exercise 6.13),

$$[L,H] = 0.$$

So L is a constant of the motion.

The Poisson brackets of K_x and K_y with H are given by

$$\begin{aligned}
[K_x,H] &= \left[p_y L - mk\frac{x}{r}, \frac{1}{2m}\left(p_x^2 + p_y^2\right) - \frac{k}{r}\right] \\
&= -kL\left[p_y, \frac{1}{r}\right] - \frac{k}{2}\left[\frac{x}{r}, p_x^2 + p_y^2\right] \\
&= kL\frac{\partial}{\partial y}\left(\frac{1}{r}\right) - kp_x\frac{\partial}{\partial x}\left(\frac{x}{r}\right) - kp_y\frac{\partial}{\partial y}\left(\frac{x}{r}\right) \\
&= k\left(xp_y - yp_x\right)\left(\frac{-y}{r^3}\right) - kp_x\left(\frac{1}{r} - \frac{x^2}{r^3}\right) - kp_y\left(\frac{-xy}{r^3}\right) \\
&= -kp_x\left(\frac{1}{r} - \frac{x^2 + y^2}{r^3}\right) = 0,
\end{aligned}$$

$$\begin{aligned}
[K_y,H] &= \left[-p_x L - mk\frac{y}{r}, \frac{1}{2m}\left(p_x^2 + p_y^2\right) - \frac{k}{r}\right] \\
&= kL\left[p_x, \frac{1}{r}\right] - \frac{k}{2}\left[\frac{y}{r}, p_x^2 + p_y^2\right] \\
&= -kL\frac{\partial}{\partial x}\left(\frac{1}{r}\right) - kp_x\frac{\partial}{\partial x}\left(\frac{y}{r}\right) - kp_y\frac{\partial}{\partial y}\left(\frac{y}{r}\right) \\
&= k\left(xp_y - yp_x\right)\left(\frac{x}{r^3}\right) + kp_x\left(\frac{xy}{r^3}\right) - kp_y\left(\frac{1}{r} - \frac{x^2}{r^3}\right) \\
&= -kp_y\left(\frac{1}{r} - \frac{x^2 + y^2}{r^3}\right) = 0.
\end{aligned}$$

So K_x and K_y are constants of the motion (see also Exercise 1.12).

(b) K_x and K_y are the x- and y-components of the three-dimensional Laplace-Runge-Lenz vector,

$$\mathbf{K} = \mathbf{p} \times \mathbf{L} - mk\,\mathbf{r}/r,$$

and L is the z-component of the three-dimensional angular momentum vector $\mathbf{L} = \mathbf{r} \times \mathbf{p}$. Exercise 6.13 shows that for any vector **K**,

$$[\mathbf{K}, \mathbf{L} \cdot \mathbf{n}] = \mathbf{n} \times \mathbf{K}.$$

Setting $\mathbf{n} = \mathbf{k}$ and taking the x- and y- components of the resulting equation, we obtain the Poisson brackets of K_x and K_y with L,

$$[K_x, L] = (\mathbf{k} \times \mathbf{K})_x = -K_y, \qquad [K_y, L] = (\mathbf{k} \times \mathbf{K})_y = K_x.$$

The Poisson bracket of K_x and K_y with each other is given by

$$\begin{aligned}
[K_x, K_y] &= \left[p_y L - mk\frac{x}{r}, -p_x L - mk\frac{y}{r}\right] \\
&= -[p_y L, p_x L] - mk\left[p_y L, \frac{y}{r}\right] + mk\left[\frac{x}{r}, p_x L\right] \\
&= -p_y L[L, p_x] - p_x L[p_y, L] - \frac{mk}{r}\left(p_y[L, y] - p_x[x, L]\right) - mkL\left(\left[p_y, \frac{y}{r}\right] - \left[\frac{x}{r}, p_x\right]\right)
\end{aligned}$$

Exercise 6.13 then shows that

$$\begin{aligned}
[K_x, K_y] &= -p_y^2 L - p_x^2 L + \frac{mk}{r}\left(p_y x - p_x y\right) + mkL\left(\frac{\partial}{\partial y}\left(\frac{y}{r}\right) + \frac{\partial}{\partial x}\left(\frac{x}{r}\right)\right) \\
&= -\left(p_x^2 + p_y^2\right)L + \frac{mk}{r}L + mkL\left(\frac{2}{r} - \frac{x^2 + y^2}{r^3}\right) \\
&= -2m\left(\frac{p_x^2 + p_y^2}{2m} - \frac{k}{r}\right)L = -2mHL.
\end{aligned}$$

For bound states $H = -|E|$ where E is the energy and we can write the preceding results in the form

$$\left[\frac{K_y}{\sqrt{2m|E|}}, L\right] = \frac{K_x}{\sqrt{2m|E|}}, \quad \left[L, \frac{K_x}{\sqrt{2m|E|}}\right] = \frac{K_y}{\sqrt{2m|E|}}, \quad \left[\frac{K_x}{\sqrt{2m|E|}}, \frac{K_y}{\sqrt{2m|E|}}\right] = L.$$

These say that the set of three quantities $\left(\frac{K_x}{\sqrt{2m|E|}}, \frac{K_y}{\sqrt{2m|E|}}, L\right)$ have the same Poisson bracket relations with one another as the components of an abstract "three-dimensional angular momentum."

(c) Now consider

$$\begin{aligned}
K_x^2 + K_y^2 &= \left(p_y L - mk\frac{x}{r}\right)^2 + \left(-p_x L - mk\frac{y}{r}\right)^2 \\
&= \left(p_y^2 + p_x^2\right)L^2 - \frac{2mk}{r}\left(xp_y - yp_x\right)L + m^2k^2\frac{x^2+y^2}{r^2} \\
&= 2m\left(\frac{p_x^2+p_y^2}{2m} - \frac{k}{r}\right)L^2 + m^2k^2 \\
&= 2mHL^2 + m^2k^2 .
\end{aligned}$$

If we restrict ourselves to bound states, for which $H = -|E|$ where E is the energy, we can rewrite this in the form

$$\left(\frac{K_x}{\sqrt{2m|E|}}\right)^2 + \left(\frac{K_y}{\sqrt{2m|E|}}\right)^2 + L^2 = \frac{mk^2}{2|E|}.$$

This says that the square of the length of the abstract "three-dimensional angular momentum" is $mk^2/2|E|$.

In quantum mechanics the set of three quantities $\left(\frac{K_x}{\sqrt{2m|E|}}, \frac{K_y}{\sqrt{2m|E|}}, L\right)$, defined in an appropriately symmetrized way, have the same Poisson bracket relations with one another as in classical mechanics, so they are again like the components of "three-dimensional angular momentum." The square of the length of this "three-dimensional angular momentum" in quantum mechanics is

$$\left(\frac{K_x}{\sqrt{2m|E|}}\right)^2 + \left(\frac{K_y}{\sqrt{2m|E|}}\right)^2 + L^2 = \frac{mk^2}{2|E|} - \frac{\hbar^2}{4}.$$

(Note the "extra" term $-\hbar^2/4$). Angular momentum theory shows that the eigenvalues of this quantity are $\lambda(\lambda+1)\hbar^2$ where $\lambda = 0,1,2\cdots$ (no half integer in this case, since the third component L really is *orbital* angular momentum) so the eigenvalues of the energy of a two-dimensional hydrogen atom are

$$E_n = -\frac{mk^2}{2\hbar^2}\frac{4}{(2n+1)^2} \qquad \text{where} \qquad n = 0,1,2,\cdots.$$

CHAPTER VII

CANONICAL TRANSFORMATIONS

Exercise 7.01

The motion of a particle of mass m undergoing constant acceleration a in one dimension is described by

$$x = x_0 + \frac{p_0}{m}t + \frac{1}{2}at^2, \qquad p = p_0 + mat.$$

Show that the transformation from present "old" variables (x, p) to initial "new" variables (x_0, p_0) is a canonical transformation
(a) by Poisson bracket test
(b) by finding (for $t \neq 0$) the type 1 generating function $F_1(x, x_0, t)$.

Solution

(a) The initial "new" variables (x_0, p_0) are given in terms of the present "old" variables (x, p) by the inverse transformation

$$x_0 = x - \frac{p}{m}t + \frac{1}{2}at^2, \qquad p_0 = p - mat.$$

The Poisson bracket of the initial variables with respect to the present variables is then

$$\left[x_0, p_0\right]_{x,p} = \frac{\partial x_0}{\partial x}\frac{\partial p_0}{\partial p} - \frac{\partial x_0}{\partial p}\frac{\partial p_0}{\partial x} = 1 \times 1 - \left(-\frac{t}{m}\right) \times 0 = 1.$$

Since it equals 1, the transformation from (x, p) to (x_0, p_0) is canonical.

(b) In a type 1 transformation the coordinates x and x_0 are taken as the independent variables; the momenta p and p_0 are expressed in terms of these,

$$p = m\frac{x - x_0}{t} + \frac{1}{2}mat, \qquad p_0 = m\frac{x - x_0}{t} - \frac{1}{2}mat.$$

Now consider the differential form

$$p\,dx - p_0\,dx_0 = \left(m\frac{x - x_0}{t} + \frac{1}{2}mat\right)dx - \left(m\frac{x - x_0}{t} - \frac{1}{2}mat\right)dx_0$$
$$= d\left(\frac{m(x - x_0)^2}{2t} + \frac{1}{2}mat(x + x_0)\right).$$

Since it is an exact differential, this again shows that the transformation is canonical. This approach also gives the type 1 generating function,

$$F_1(x, x_0, t) = \frac{m(x - x_0)^2}{2t} + \frac{1}{2}mat(x + x_0) + f(t)$$

where $f(t)$ is an arbitrary function of time.

Exercise 7.02

(a) Show that

$$Q = -p, \qquad P = q + Ap^2.$$

(where A is any constant) is a canonical transformation,

(i) by evaluating $[Q, P]_{q,p}$

(ii) by expressing $p\,dq - P\,dQ$ as an exact differential $dF(q, Q)$. Hence find the type 1 generating function of the transformation. To do this, you must first use the transformation equations to express p, P in terms of q, Q.

(b) Use the relation $F_2 = F_1 + PQ$ to find the type 2 generating function $F_2(q, P)$, and check your result by showing that F_2 indeed generates the transformation.

Solution

(a) (i) The Poisson bracket of the new variables with respect to the old is

$$[Q, P]_{q,p} = \frac{\partial Q}{\partial q}\frac{\partial P}{\partial p} - \frac{\partial Q}{\partial p}\frac{\partial P}{\partial q} = 0 \times (2Ap) - (-1)(1) = 1.$$

Since it equals 1, the transformation is canonical.

(ii) Suppose that we regard this as a type 1 transformation, taking the coordinates q and Q as the independent variables. The momenta p and P are then given by

$$p = -Q, \qquad P = q + AQ^2.$$

Now consider the differential form

$$p\,dq - P\,dQ = (-Q)\,dq - (q + AQ^2)\,dQ = d\left(-qQ - \tfrac{1}{3}AQ^3\right).$$

Since it is an exact differential, this again shows that the transformation is canonical. The type 1 generating function is

$$F_1(q,Q) = -qQ - \tfrac{1}{3}AQ^3.$$

(b) Suppose, instead, that we regard this as a type 2 transformation, taking the old coordinate q and the new momentum P as the independent variables. We can find the type 2 generating function from the type 1 generating function by setting

$$F_2(q,P) = F_1 + PQ = -qQ - \tfrac{1}{3}AQ^3 + (q + AQ^2)Q = \tfrac{2}{3}AQ^3 = \frac{2}{3}A\left(\frac{P-q}{A}\right)^{3/2}.$$

To check that the final expression is indeed the type 2 generating function, we evaluate its derivatives,

$$\left(\frac{\partial F_2}{\partial q}\right)_P = -\frac{1}{\sqrt{A}}(P-q)^{1/2} = p \quad \text{and} \quad \left(\frac{\partial F_2}{\partial P}\right)_q = \frac{1}{\sqrt{A}}(P-q)^{1/2} = Q.$$

These are as required.

Exercise 7.03

The Hamiltonian for a particle moving vertically in a uniform gravitational field g is

$$H = \frac{p^2}{2m} + mgq.$$

(a) Find the new Hamiltonian for new canonical variables Q, P as given in Exercise 7.02. Show that we can eliminate Q from the Hamiltonian (make Q cyclic) by choosing the constant A appropriately.
(b) With this choice of A write down and solve Hamilton's equations for the new canonical variables, and then use the transformation equations to find the original variables q, p as functions of time.

Solution

(a) We introduce new canonical variables Q and P by setting, as in Exercise 7.02,

$$q = P - AQ^2, \qquad p = -Q.$$

The Hamiltonian for the new canonical variables is

$$K(Q,P) = H(q(Q,P),p(Q,P)) = \frac{(-Q)^2}{2m} + mg(P - AQ^2) = mgP + \left(\frac{1}{2m} - mgA\right)Q^2.$$

If we take

$$A = \frac{1}{2m^2 g},$$

we can simplify this to

$$K = mgP.$$

(b) Hamilton's equations for the new canonical variables (Q, P) are then

$$\frac{dQ}{dt} = \frac{\partial K}{\partial P} = mg, \qquad \frac{dP}{dt} = -\frac{\partial K}{\partial Q} = 0,$$

with solution

$$Q = Q_0 + mgt, \qquad P = P_0.$$

Here Q_0 and P_0 are constants, the initial values of the new variables. The equations of the canonical transformation then give the original variables as functions of time,

$$q(t) = P_0 - \frac{1}{2m^2 g}(Q_0 + mgt)^2 = \left(P_0 - \frac{Q_0^2}{2m^2 g}\right) - \frac{Q_0}{m}t - \frac{1}{2}gt^2, \qquad p(t) = -Q_0 - mgt.$$

This is the well-known solution to the free fall problem. We see that $-Q_0$ is the initial momentum and P_0 is the maximum height reached.

Exercise 7.04

(a) Show that

$$Q = q\cos\theta - \frac{p}{m\omega}\sin\theta, \qquad P = m\omega q\sin\theta + p\cos\theta,$$

is a canonical transformation,

(i) by evaluating $[Q, P]_{q,p}$

(ii) by expressing $p\,dq - P\,dQ$ as an exact differential $dF_1(q, Q, t)$. Hence find the type 1 generating function of the transformation. To do this, you must first use the transformation equations to express p, P in terms of q, Q.

(b) Use the relation $F_2 = F_1 + PQ$ to find the type 2 generating function $F_2(q, P)$, and check your result by showing that F_2 indeed generates the transformation.

Solution

(a) (i) The Poisson bracket of the new variables with respect to the old is

$$[Q,P]_{q,p} = \frac{\partial Q}{\partial q}\frac{\partial P}{\partial p} - \frac{\partial Q}{\partial p}\frac{\partial P}{\partial q} = \cos\theta \times \cos\theta - \left(-\frac{1}{m\omega}\sin\theta\right)(m\omega\sin\theta) = 1.$$

Since it equals 1, the transformation from (q,p) to (Q,P) is canonical.

(ii) Suppose that we regard this as a type 1 transformation, taking the coordinates q and Q as the independent variables. The momenta p and P are then given by

$$p = m\omega\left(q\cot\theta - \frac{Q}{\sin\theta}\right), \qquad P = m\omega\left(\frac{q}{\sin\theta} - Q\cot\theta\right).$$

Now consider the differential form

$$\begin{aligned} p\,dq - P\,dQ &= m\omega\left(q\cot\theta - \frac{Q}{\sin\theta}\right)dq - m\omega\left(\frac{q}{\sin\theta} - Q\cot\theta\right)dQ \\ &= d\left(\frac{1}{2}m\omega(q^2+Q^2)\cot\theta - m\omega\frac{qQ}{\sin\theta}\right). \end{aligned}$$

Since it is an exact differential, this again shows that the transformation is canonical. The type 1 generating function is

$$F_1(q,Q) = \frac{1}{2}m\omega(q^2+Q^2)\cot\theta - m\omega\frac{qQ}{\sin\theta}.$$

(b) The type 2 generating function $F_2(q,P)$ can be obtained by setting

$$\begin{aligned} F_2 &= F_1 + PQ \\ &= \frac{1}{2}m\omega(q^2+Q^2)\cot\theta - m\omega\frac{qQ}{\sin\theta} + m\omega\left(\frac{q}{\sin\theta} - Q\cot\theta\right)Q \\ &= \frac{1}{2}m\omega(q^2-Q^2)\cot\theta. \end{aligned}$$

We must still express this in terms of the appropriate type 2 variables, q and P, by setting

$$Q = \frac{q}{\cos\theta} - \frac{P}{m\omega}\tan\theta.$$

This gives

$$F_2(q,P) = \frac{1}{2}m\omega\left[q^2 - \left(\frac{q}{\cos\theta} - \frac{P}{m\omega}\tan\theta\right)^2\right]\cot\theta$$

$$= \frac{qP}{\cos\theta} - \frac{1}{2}m\omega\left(q^2 + \frac{P^2}{m^2\omega^2}\right)\tan\theta .$$

To check this expression, we evaluate its derivatives

$$\left(\frac{\partial F_2}{\partial q}\right)_P = \frac{P}{\cos\theta} - m\omega q\tan\theta = p, \qquad \left(\frac{\partial F_2}{\partial P}\right)_q = \frac{q}{\cos\theta} - \frac{P}{m\omega}\tan\theta = Q.$$

These are as required.

Exercise 7.05

Suppose that the (q, p) of Exercise 7.04 are canonical variables for a simple harmonic oscillator with Hamiltonian

$$H = \frac{p^2}{2m} + \frac{1}{2}m\omega^2 q^2.$$

(a) Find the Hamiltonian $K(Q,P,t)$ for the new canonical variables (Q,P), assuming that the parameter θ is some function of time. Show that we can choose $\theta(t)$ so that $K = 0$.
(b) With this choice of $\theta(t)$ solve the new canonical equations to find (Q,P) as functions of time, and then use the transformation equations to find the original variables (q,p) as functions of time.

Solution

(a) We introduce new canonical variables (Q,P) by setting, as in Exercise 7.04,

$$q = Q\cos\theta + \frac{P}{m\omega}\sin\theta, \qquad p = -m\omega Q\sin\theta + P\cos\theta.$$

The Hamiltonian $K(Q,P,t)$ for the new canonical variables is given by

$$K(Q,P,t) = H(q,p) + \left(\frac{\partial F_2}{\partial t}\right)_{q,P}$$

where

$$F_2(q,P,t) = \frac{qP}{\cos\theta} - \frac{1}{2}m\omega\left(q^2 + \frac{P^2}{m^2\omega^2}\right)\tan\theta$$

is the type 2 generating function of the transformation. The first term in $K(Q,P,t)$ is the old Hamiltonian $H(q,p)$, expressed in terms of the new variables,

$$\begin{aligned} H(q,p) &= \frac{1}{2m}p^2 + \frac{1}{2}m\omega^2 q^2 \\ &= \frac{1}{2m}(-m\omega Q\sin\theta + P\cos\theta)^2 + \frac{1}{2}m\omega^2\left(Q\cos\theta + \frac{P}{m\omega}\sin\theta\right)^2 \\ &= \frac{1}{2m}P^2 + \frac{1}{2}m\omega^2 Q^2 = H(Q,P)\,. \end{aligned}$$

The second term in $K(Q,P,t)$ is the time derivative of the generating function,

$$\begin{aligned} \left(\frac{\partial F_2}{\partial t}\right)_{q,P} &= \left[qP\sin\theta - \frac{1}{2}m\omega\left(q^2 + \frac{P^2}{m^2\omega^2}\right)\right]\frac{\dot\theta}{\cos^2\theta} \\ &= \left[\left(Q\cos\theta + \frac{P}{m\omega}\sin\theta\right)P\sin\theta - \frac{1}{2}m\omega\left(\left(Q\cos\theta + \frac{P}{m\omega}\sin\theta\right)^2 + \frac{P^2}{m^2\omega^2}\right)\right]\frac{\dot\theta}{\cos^2\theta} \\ &= -\left(\frac{P^2}{2m\omega} + \frac{1}{2}m\omega Q^2\right)\dot\theta = -H(Q,P)(\dot\theta/\omega)\,. \end{aligned}$$

The new Hamiltonian $K(Q,P,t)$ is thus

$$K(Q,P,t) = H(Q,P)(1 - \dot\theta/\omega)$$

and reduces to zero if we take $\theta = \omega t$.

(b) Hamilton's equations for the new variables are then

$$\frac{dQ}{dt} = \frac{\partial K}{\partial P} = 0, \qquad \frac{dP}{dt} = -\frac{\partial K}{\partial Q} = 0,$$

so the new canonical variables are constants $Q = Q_0$, $P = P_0$. The equations of the canonical transformation then give the original variables (q,p) as functions of time,

$$q(t) = Q_0\cos\omega t + \frac{P_0}{m\omega}\sin\omega t, \qquad p(t) = -m\omega Q_0\sin\omega t + P_0\cos\omega t\,.$$

This is the well-known solution to the harmonic oscillator problem. The new canonical variables (Q_0,P_0) are the initial $(t = 0)$ values of the original variables (q,p).

Exercise 7.06

(a) Show that the Hamiltonian for a simple harmonic oscillator is invariant under the canonical transformation of Exercise 7.04 (for θ constant).
(b) Find the associated constant of the motion.

Solution

(a) We showed in Exercise 7.05 that under the canonical transformation

$$q = Q\cos\theta + (P/m\omega)\sin\theta, \qquad p = -m\omega Q\sin\theta + P\cos\theta,$$

the Hamiltonian of a simple harmonic oscillator,

$$H(q,p) = \frac{p^2}{2m} + \frac{1}{2}m\omega^2 q^2,$$

becomes (for θ constant)

$$K(Q,P) = \frac{P^2}{2m} + \frac{1}{2}m\omega^2 Q^2 = H(Q,P).$$

The new Hamiltonian has the same form as the old and thus the system is invariant under this transformation. This also follows from the observation that the transformation is a rotation of phase space $(q, p/m\omega)$ through an angle θ. Under such rotation the $(\text{distance})^2$ from the origin, $q^2 + (p/m\omega)^2 = 2H/m\omega^2$, is invariant.

(b) To find the associated constant of the motion, we consider the infinitesimal transformation obtained by replacing θ by the infinitesimal $\delta\theta$. The type 2 generating function becomes

$$F_2(q,P,t) = \frac{qP}{\cos\theta} - \frac{m\omega}{2}\left(q^2 + \frac{P^2}{m^2\omega^2}\right)\tan\theta \approx qP - \frac{m\omega}{2}\left(q^2 + \frac{P^2}{m^2\omega^2}\right)\delta\theta.$$

The first term in F_2 is the generating function qP of the identity transformation and the second is the generator εG of the infinitesimal transformation,

$$\varepsilon G(q,p,t) = -\frac{m\omega}{2}\left(q^2 + \frac{p^2}{m^2\omega^2}\right)\delta\theta = -(H(q,p)/\omega)\delta\theta.$$

This generator, proportional to the action variable $I = H/\omega$, is the constant of the motion associated with this invariance transformation.

Exercise 7.07

(a) What are the conditions on the "small" constants a, b, c, d, e, and f in order that

$$q = Q + aQ^2 + 2bQP + cP^2$$
$$p = P + dQ^2 + 2eQP + fP^2$$

be a canonical transformation to first order in small quantities?
(b) The Hamiltonian for a slightly anharmonic oscillator is

$$H = \frac{p^2}{2m} + \frac{1}{2}m\omega^2 q^2 + \beta q^3$$

where β is "small." Perform a canonical transformation of the type given in part (a) and adjust the constants so that the new Hamiltonian does not contain an anharmonic term to first order in small quantities, thus

$$K = \frac{P^2}{2m} + \frac{1}{2}m\omega^2 Q^2 + \text{second order terms.}$$

(c) Write down and solve Hamilton's equations for the new canonical variables, and then use the transformation equations to find the solution to the anharmonic oscillator problem valid to first order in small quantities.

Solution

(a) To determine the conditions under which this transformation is canonical, we evaluate the Poisson bracket

$$\begin{aligned}[q,p]_{Q,P} &= \frac{\partial q}{\partial Q}\frac{\partial p}{\partial P} - \frac{\partial q}{\partial P}\frac{\partial p}{\partial Q} \\ &= (1 + 2aQ + 2bP)(1 + 2eQ + 2fP) - (2bQ + 2cP)(2dQ + 2eP) \\ &= 1 + 2(a + e)Q + 2(b + f)P + \text{second order terms}.\end{aligned}$$

For a canonical transformation this must equal 1, and so to first order in small quantities we require

$$a + e = 0 \quad \text{and} \quad b + f = 0.$$

(b) Now consider a slightly anharmonic oscillator with Hamiltonian

$$H(q,p) = \frac{p^2}{2m} + \frac{1}{2}m\omega^2 q^2 + \beta q^3$$

where β is "small." We transform from old variables (q,p) to new variables (Q,P) by using the canonical transformation of part (a). The Hamiltonian for the new variables is

$$
\begin{aligned}
K(Q,P) &= H(q(Q,P),p(Q,P)) \\
&= \frac{1}{2m}\left(P + dQ^2 + 2eQP + fP^2\right)^2 + \frac{1}{2}m\omega^2\left(Q + aQ^2 + 2bQP + cP^2\right)^2 \\
&\quad +\beta\left(Q + aQ^2 + 2bQP + cP^2\right)^3 \\
&= \frac{1}{2m}P^2 + \frac{1}{2}m\omega^2Q^2 \\
&\quad +\left(\frac{d}{m} + 2bm\omega^2\right)PQ^2 + \left(\frac{2e}{m} + cm\omega^2\right)QP^2 + \frac{f}{m}P^3 + \left(am\omega^2 + \beta\right)Q^3 + \cdots .
\end{aligned}
$$

We choose the constants so that K does not contain an anharmonic term to first order in small quantities,

$$
\frac{d}{m} + 2bm\omega^2 = 0, \qquad \frac{2e}{m} + cm\omega^2 = 0, \qquad \frac{f}{m} = 0, \qquad am\omega^2 + \beta = 0.
$$

Combining these conditions with the requirement that the transformation be canonical, we find

$$
a = -\frac{\beta}{m\omega^2}, \quad c = -\frac{2\beta}{m^3\omega^4}, \quad e = \frac{\beta}{m\omega^2}, \quad b = 0, \quad d = 0, \quad f = 0.
$$

The transformation becomes

$$
q = Q - \frac{\beta}{m\omega^2}\left(Q^2 + \frac{2P^2}{m^2\omega^2}\right), \qquad p = P + \frac{2\beta}{m\omega^2}QP.
$$

(c) Hamilton's equations for the new variables are then

$$
\frac{dQ}{dt} = \frac{\partial K}{\partial P} = \frac{P}{m} + \cdots, \qquad \frac{dP}{dt} = -\frac{\partial K}{\partial Q} = -m\omega^2Q + \cdots .
$$

These are the equations of motion of a simple harmonic oscillator, with solution

$$
Q(t) = A\sin(\omega t + \phi), \qquad P(t) = m\omega A\cos(\omega t + \phi),
$$

where A and ϕ are constants. The solution to the anharmonic oscillator problem, valid to first order in small quantities, is then given by the equations of the canonical transformation,

$$q(t) = A\sin(\omega t + \phi) - \frac{\beta A^2}{m\omega^2}\left(\sin^2(\omega t + \phi) + 2\cos^2(\omega t + \phi)\right)$$
$$= A\sin(\omega t + \phi) - \frac{\beta A^2}{2m\omega^2}(3 + \cos 2(\omega t + \phi)),$$
$$p(t) = m\omega A\cos(\omega t + \phi) + \frac{2\beta A^2}{\omega}\sin(\omega t + \phi)\cos(\omega t + \phi)$$
$$= m\omega A\cos(\omega t + \phi) + \frac{\beta A^2}{\omega}\sin 2(\omega t + \phi) \quad \left(= m\frac{dq(t)}{dt}\right).$$

Exercise 7.08

(a) Show that

$$x' = x + \frac{1}{m}p_z\tau, \qquad p'_x = p_x - mg\tau,$$
$$y' = y, \qquad p'_y = p_y,$$
$$z' = z + \frac{1}{m}p_x\tau - \frac{1}{2}g\tau^2, \qquad p'_z = p_z,$$

(where τ is any constant) is a canonical transformation by finding the type 2 generating function $F_2(x,y,z,p'_x,p'_y,p'_z)$.

(b) Show that the Hamiltonian

$$H = \frac{1}{2m}(p_x^2 + p_y^2 + p_z^2) + mgz$$

for a projectile near the surface of the earth is invariant under the canonical transformation given in part (a), and find the associated constant of the motion.

Solution

(a) To view this as a type 2 transformation, we take the old coordinates (x,y,z) and the new momenta (p'_x,p'_y,p'_z) as the independent variables and express the new coordinates (x',y',z') and old momenta (p_x,p_y,p_z) in terms of them,

$$x' = x + (\tau/m)p'_z, \qquad p_x = p'_x + mg\tau,$$
$$y' = y, \qquad p_y = p'_y,$$
$$z' = z + (\tau/m)p'_x + (g\tau^2/2), \qquad p_z = p'_z.$$

Now consider the differential form

$$p_x\,dx + p_y\,dy + p_z\,dz + x'\,dp'_x + y'\,dp'_y + z'\,dp'_z$$
$$= (p'_x + mg\tau)dx + p'_y\,dy + p'_z\,dz + \left(x + (\tau/m)p'_z\right)dp'_x + y\,dp'_y + \left(z + (\tau/m)p'_x + (g\tau^2/2)\right)dp'_z$$
$$= d\left(xp'_x + yp'_y + zp'_z + mg\tau x + (\tau/m)p'_x p'_z + (g\tau^2/2)p'_z\right).$$

Since it is an exact differential, the transformation is canonical with type 2 generating function

$$F_2(x,y,z,p'_x,p'_y,p'_z) = xp'_x + yp'_y + zp'_z + mg\tau x + (\tau/m)p'_x p'_z + (g\tau^2/2)p'_z.$$

(b) Consider the Hamiltonian

$$H(q,p) = \frac{1}{2m}\left(p_x^2 + p_y^2 + p_z^2\right) + mgz$$

for projectile motion near the surface of the earth and introduce new variables by using the canonical transformation of part (a). The Hamiltonian for these new variables is given by

$$\begin{aligned} K(Q,P) &= H\big(q(Q,P), p(Q,P)\big) \\ &= \frac{1}{2m}\left((p'_x + mg\tau)^2 + p'^2_y + p'^2_z\right) + mg\left(z' - (\tau/m)p'_x - (g\tau^2/2)\right) \\ &= \frac{1}{2m}\left(p'^2_x + p'^2_y + p'^2_z\right) + mgz' = H(Q,P). \end{aligned}$$

The Hamiltonian, and hence the system, is invariant under the transformation. The associated infinitesimal transformation is generated by

$$F_2 = \left(xp'_x + yp'_y + zp'_z\right) + \left(mgx + p'_x p'_z/m\right)\delta\tau.$$

The first term in F_2 is the generating function of the identity transformation and the second is the generator εG of the infinitesimal transformation,

$$\varepsilon G(q,p) = \left(mgx + p_x p_z/m\right)\delta\tau.$$

This generator, $mgx + p_x p_z/m$, is the constant of the motion associated with this infinitesimal invariance transformation. To check that this quantity is indeed a constant of the motion, we evaluate its time derivative,

$$\frac{d}{dt}\left(mgx + \frac{1}{m}p_x p_z\right) = mg\frac{dx}{dt} + \frac{p_x}{m}\frac{dp_z}{dt} + \frac{p_z}{m}\frac{dp_x}{dt} = \frac{p_x}{m}\left(mg + \frac{dp_z}{dt}\right) + \frac{p_z}{m}\frac{dp_x}{dt} = 0.$$

Here we have made use of the well-known properties of projectile motion, $dp_z/dt = -mg$ and $dp_x/dt = 0$.

Exercise 7.09

(a) Show that

$$Q_1 = \frac{1}{\sqrt{2}}\left(q_1 + \frac{p_2}{m\omega}\right), \quad P_1 = \frac{1}{\sqrt{2}}(p_1 - m\omega q_2),$$
$$Q_2 = \frac{1}{\sqrt{2}}\left(q_1 - \frac{p_2}{m\omega}\right), \quad P_2 = \frac{1}{\sqrt{2}}(p_1 + m\omega q_2),$$

(where $m\omega$ is a constant) is a canonical transformation by Poisson bracket test. This requires evaluating *six* simple Poisson brackets.
(b) Find a generating function $F(q_1,q_2,Q_1,P_2)$ for this transformation, type 1 in the first degree of freedom and type 2 in the second degree of freedom.

Solution

(a) To check whether this is a canonical transformation, we evaluate the fundamental Poisson brackets,

$$[Q_1,Q_2] = 0, \qquad [Q_2,P_1] = \tfrac{1}{2} - \tfrac{1}{2} = 0,$$
$$[Q_1,P_1] = \tfrac{1}{2} + \tfrac{1}{2} = 1, \qquad [Q_2,P_2] = \tfrac{1}{2} + \tfrac{1}{2} = 1,$$
$$[Q_1,P_2] = \tfrac{1}{2} - \tfrac{1}{2} = 0, \qquad [P_1,P_2] = 0.$$

These are as required, so the transformation is indeed canonical.

(b) We now wish to find a generating function for this transformation. We *cannot* regard the transformation as a pure type 1 transformation with q_1,q_2,Q_1,Q_2 independent, since the equations imply that

$$Q_1 + Q_2 = \sqrt{2}\,q_1,$$

nor can we regard it as a pure type 2 transformation with q_1,q_2,P_1,P_2 independent, since the equations imply that

$$-P_1 + P_2 = \sqrt{2}\,m\omega q_2.$$

We can, however, regard it as type 1 in the first degree of freedom and type 2 in the second degree of freedom with q_1,q_2,Q_1,P_2 independent, since we *can* express the other variables in terms of these,

$$p_1 = \sqrt{2}\,P_2 - m\omega q_2, \qquad P_1 = P_2 - \sqrt{2}\,m\omega q_2,$$
$$p_2 = m\omega\left(\sqrt{2}\,Q_1 - q_1\right), \qquad Q_2 = \sqrt{2}\,q_1 - Q_1.$$

Now consider the differential form

$$p_1\,dq_1 + p_2\,dq_2 - P_1\,dQ_1 + Q_2\,dP_2 = \left(\sqrt{2}\,P_2 - m\omega q_2\right)dq_1 + m\omega\left(\sqrt{2}\,Q_1 - q_1\right)dq_2$$
$$-\left(P_2 - \sqrt{2}\,m\omega q_2\right)dQ_1 + \left(\sqrt{2}q_1 - Q_1\right)dP_2$$
$$= d\left(\sqrt{2}\,q_1P_2 - m\omega q_1q_2 + \sqrt{2}\,m\omega q_2Q_1 - Q_1P_2\right).$$

Since it is an exact differential, this again shows that the transformation is canonical. The mixed-type generating function is

$$F(q_1,q_2,Q_1,P_2) = \sqrt{2}\,q_1P_2 - m\omega q_1q_2 + \sqrt{2}\,m\omega q_2Q_1 - Q_1P_2.$$

Note that the inverse transformation

$$q_1 = \frac{1}{\sqrt{2}}(Q_1 + Q_2), \qquad p_1 = \frac{1}{\sqrt{2}}(P_1 + P_2),$$
$$q_2 = \frac{1}{\sqrt{2}\,m\omega}(-P_1 + P_2), \qquad p_2 = \frac{m\omega}{\sqrt{2}}(Q_1 - Q_2),$$

enables us to set

$$\omega\left(q_1p_2 - q_2p_1\right) = \left(\frac{P_1^2}{2m} + \frac{1}{2}m\omega^2Q_1^2\right) - \left(\frac{P_2^2}{2m} + \frac{1}{2}m\omega^2Q_2^2\right).$$

That is, it enables us to express the "third component of angular momentum" as the difference of two "harmonic oscillator Hamiltonians." In quantum mechanics this can be used to show that a component of *orbital* angular momentum has eigenvalues $n\hbar$ where n is an *integer* (and not a half integer). See Leslie E. Ballentine, *Quantum Mechanics: A Modern Development* (World Scientific Pub. Co. Pte. Ltd., Singapore, 1998), p. 169.

Exercise 7.10

(a) Let x denote a column matrix of the canonical variables q_1,q_2,p_1,p_2 for a system with two degrees of freedom, and consider a linear transformation

$$x \rightarrow x' = Mx$$

where M is a 4×4 matrix with constant elements. Use the Poisson bracket conditions to find the conditions on the elements of M in order that this be a canonical transformation.
(b) Show that these are equivalent to requiring that M satisfy the condition

$$MJ\tilde{M} = J.$$

Here $\tilde{M}$ is the transpose of M, and J is the matrix

$$J = \begin{bmatrix} 0 & 1 \\ -1 & 0 \end{bmatrix}$$

where "0" stands for the 2×2 zero matrix and "1" stands for the 2×2 unit matrix. Matrices M which satisfy the above condition are called symplectic matrices.

Solution

(a) Consider a linear transformation of the canonical variables,

$$\begin{bmatrix} Q_1 \\ Q_2 \\ P_1 \\ P_2 \end{bmatrix} = \begin{bmatrix} a_{11} & a_{12} & b_{11} & b_{12} \\ a_{21} & a_{22} & b_{21} & b_{22} \\ c_{11} & c_{12} & d_{11} & d_{12} \\ c_{21} & c_{22} & d_{21} & d_{22} \end{bmatrix} \begin{bmatrix} q_1 \\ q_2 \\ p_1 \\ p_2 \end{bmatrix}.$$

We want to find the conditions on the constant coefficients a, b, c, and d in order that this be a canonical transformation. These are given by the conditions on the fundamental Poisson brackets of the new variables, namely

$$\begin{aligned}
[Q_1, Q_2] &= a_{11}b_{21} - b_{11}a_{21} + a_{12}b_{22} - b_{12}a_{22} = 0, \\
[Q_1, P_1] &= a_{11}d_{11} - b_{11}c_{11} + a_{12}d_{12} - b_{12}c_{12} = 1, \\
[Q_1, P_2] &= a_{11}d_{21} - b_{11}c_{21} + a_{12}d_{22} - b_{12}c_{22} = 0, \\
[Q_2, P_1] &= a_{21}d_{11} - b_{21}c_{11} + a_{22}d_{12} - b_{22}c_{12} = 0, \\
[Q_2, P_2] &= a_{21}d_{21} - b_{21}c_{21} + a_{22}d_{22} - b_{22}c_{22} = 1, \\
[P_1, P_2] &= c_{11}d_{21} - d_{11}c_{21} + c_{12}d_{22} - d_{12}c_{22} = 0;
\end{aligned}$$

that is, six conditions on the sixteen coefficients.

(b) Let M be the matrix of the coefficients,

$$M = \begin{bmatrix} a_{11} & a_{12} & b_{11} & b_{12} \\ a_{21} & a_{22} & b_{21} & b_{22} \\ c_{11} & c_{12} & d_{11} & d_{12} \\ c_{21} & c_{22} & d_{21} & d_{22} \end{bmatrix},$$

and let J be the 4×4 matrix

$$J = \begin{bmatrix} 0 & 0 & 1 & 0 \\ 0 & 0 & 0 & 1 \\ -1 & 0 & 0 & 0 \\ 0 & -1 & 0 & 0 \end{bmatrix}.$$

We work out

$$M J \tilde{M} = \begin{bmatrix} 0 & [Q_1,Q_2] & [Q_1,P_1] & [Q_1,P_2] \\ [Q_2,Q_1] & 0 & [Q_2,P_1] & [Q_2,P_2] \\ [P_1,Q_1] & [P_1,Q_2] & 0 & [P_1,P_2] \\ [P_2,Q_1] & [P_2,Q_2] & [P_2,P_1] & 0 \end{bmatrix}$$

where $\tilde{M}$ is the transpose of M. If the transformation is canonical, the fundamental Poisson brackets have the values given in part (a) and this becomes

$$M J \tilde{M} = J\,.$$

Conversely, if this condition holds, the fundamental Poisson brackets are as in part (a) and the transformation is canonical. For further details see Herbert Goldstein, *Classical Mechanics*, (Addison-Wesley Publishing Company, Reading, 1980) 2nd ed., sect. 9.3.

Exercise 7.11

The dynamics of a system of interacting particles is governed by a Hamiltonian

$$H = \sum_{i=1}^{N} \frac{|\mathbf{p}_i|^2}{2m_i} + \frac{1}{2}\sum_{i=1}^{N}\sum_{j=1}^{N} V_{ij}(\mathbf{r}_i - \mathbf{r}_j)\,.$$

Suppose we view this system from a uniformly accelerating coordinate frame

$$\mathbf{r}_i' = \mathbf{r}_i - \tfrac{1}{2}\mathbf{a}t^2\,.$$

Show that we can choose the canonical transformation connecting the two frames (that is, its type 2 generating function $F_2(\mathbf{r},\mathbf{p}',t)$) so that the Hamiltonian H' in the accelerating coordinate frame has the same form as H, except for an additional term which can be interpreted as arising from the presence of an effective gravitational field $-\mathbf{a}$. What is then the relation between the momenta $\mathbf{p}_i$ and $\mathbf{p}_i'$ in the two frames?

Solution

Suppose that the transformation from inertial frame coordinates $\mathbf{r}_i$ to accelerating frame coordinates $\mathbf{r}_i'$ is generated by a type 2 generating function $F_2(\mathbf{r},\mathbf{p}',t)$. Half of the transformation equations reads

$$\mathbf{r}_i' = \frac{\partial F_2}{\partial \mathbf{p}_i'} = \mathbf{r}_i - \tfrac{1}{2}\mathbf{a}t^2\,,$$

so the generating function must have the form

$$F_2 = \sum_{i=1}^{N}\left(\mathbf{r}_i - \tfrac{1}{2}\mathbf{a}t^2\right)\cdot \mathbf{p}_i' + f(\mathbf{r},t)$$

where f is a function, as yet undetermined, of the inertial coordinates and the time. The other half of the transformation equations then reads

$$\mathbf{p}_i = \frac{\partial F_2}{\partial \mathbf{r}_i} = \mathbf{p}_i' + \frac{\partial f}{\partial \mathbf{r}_i}.$$

Now suppose that the dynamics of this system, in the inertial frame, is governed by a Hamiltonian of the form

$$H = \sum_{i=1}^{N}\frac{|\mathbf{p}_i|^2}{2m_i} + \frac{1}{2}\sum_{i=1}^{N}\sum_{j=1}^{N} V_{ij}(\mathbf{r}_i - \mathbf{r}_j).$$

The Hamiltonian, in the accelerated frame, is given by

$$K(\mathbf{r}',\mathbf{p}',t) = H(\mathbf{r},\mathbf{p}) + \frac{\partial F_2}{\partial t}.$$

The first term in K is the inertial frame Hamiltonian, written in terms of the accelerating frame variables,

$$\begin{aligned}
H(\mathbf{r},\mathbf{p}) &= \sum_{i=1}^{N}\frac{1}{2m_i}\left|\mathbf{p}_i' + \frac{\partial f}{\partial \mathbf{r}_i}\right|^2 + \frac{1}{2}\sum_{i=1}^{N}\sum_{j=1}^{N} V_{ij}\left(\mathbf{r}_i' - \tfrac{1}{2}\mathbf{a}t^2 - \mathbf{r}_j' + \tfrac{1}{2}\mathbf{a}t^2\right)\\
&= \sum_{i=1}^{N}\frac{|\mathbf{p}_i'|^2}{2m_i} + \sum_{i=1}^{N}\frac{\mathbf{p}_i'}{m_i}\cdot\frac{\partial f}{\partial \mathbf{r}_i} + \sum_{i=1}^{N}\frac{1}{2m_i}\left|\frac{\partial f}{\partial \mathbf{r}_i}\right|^2 + \frac{1}{2}\sum_{i=1}^{N}\sum_{j=1}^{N} V_{ij}\left(\mathbf{r}_i' - \mathbf{r}_j'\right)\\
&= H(\mathbf{r}',\mathbf{p}') + \sum_{i=1}^{N}\frac{\mathbf{p}_i'}{m_i}\cdot\frac{\partial f}{\partial \mathbf{r}_i} + \sum_{i=1}^{N}\frac{1}{2m_i}\left|\frac{\partial f}{\partial \mathbf{r}_i}\right|^2 .
\end{aligned}$$

We should really express the remaining **r**'s in this in terms of the **r**' 's, but shall leave this for later. The second term in K is the partial time derivative of the generating function

$$\left(\frac{\partial F_2}{\partial t}\right)_{\mathbf{r},\mathbf{p}'} = -\sum_{i=1}^{N}\mathbf{p}_i'\cdot\mathbf{a}t + \frac{\partial f}{\partial t}.$$

The Hamiltonian in the accelerating frame is thus

$$K = H(\mathbf{r}',\mathbf{p}') + \sum_{i=1}^{N} \frac{\mathbf{p}_i'}{m_i} \cdot \left(\frac{\partial f}{\partial \mathbf{r}_i} - m_i \mathbf{a}t \right) + \sum_{i=1}^{N} \frac{1}{2m_i} \left| \frac{\partial f}{\partial \mathbf{r}_i} \right|^2 + \frac{\partial f}{\partial t}.$$

We choose f so as to eliminate the term linear in $\mathbf{p}_i'$. This requires

$$\frac{\partial f}{\partial \mathbf{r}_i} = m_i \mathbf{a}t, \quad \text{so} \quad f = \sum_{i=1}^{N} m_i \mathbf{r}_i \cdot \mathbf{a}t + g(t)$$

where g is a function of time, as yet undetermined. The Hamiltonian in the accelerating frame becomes

$$\begin{aligned} K &= H(\mathbf{r}',\mathbf{p}') + \sum_{i=1}^{N} \frac{1}{2} m_i |\mathbf{a}|^2 t^2 + \sum_{i=1}^{N} m_i \mathbf{r}_i \cdot \mathbf{a} + \frac{dg}{dt} \\ &= H(\mathbf{r}',\mathbf{p}') + \sum_{i=1}^{N} m_i |\mathbf{a}|^2 t^2 + \sum_{i=1}^{N} m_i \mathbf{r}_i' \cdot \mathbf{a} + \frac{dg}{dt} \end{aligned}$$

where in the second equality we have finally expressed the remaining $\mathbf{r}$'s in terms of the $\mathbf{r}'$'s. We now choose g so as to eliminate the explicit time dependence. This requires

$$\frac{dg}{dt} = -\sum_{i=1}^{N} m_i |\mathbf{a}|^2 t^2, \quad \text{so} \quad g = -\frac{1}{3} \sum_{i=1}^{N} m_i |\mathbf{a}|^2 t^3.$$

The full generating function of the transformation from an inertial frame to a uniformly accelerating frame (with our choices) is then

$$F_2 = \sum_{i=1}^{N} \left(\mathbf{r}_i - \tfrac{1}{2}\mathbf{a}t^2 \right) \cdot \mathbf{p}_i' + \sum_{i=1}^{N} m_i \mathbf{r}_i \cdot \mathbf{a}t - \frac{1}{3} \sum_{i=1}^{N} m_i |\mathbf{a}|^2 t^3.$$

The relation between the momenta in the two frames becomes

$$\mathbf{p}_i = \frac{\partial F_2}{\partial \mathbf{r}_i} = \mathbf{p}_i' + m_i \mathbf{a}t, \quad \text{so} \quad \mathbf{p}_i' = \mathbf{p}_i - m_i \mathbf{v}$$

where $\mathbf{v} = \mathbf{a}t$ is the velocity of the accelerated frame with respect to the inertial frame. This is the connection between the momenta which we would expect. The Hamiltonian in the accelerated frame becomes

$$K = H(\mathbf{r}',\mathbf{p}') + \sum_{i=1}^{N} m_i \mathbf{r}_i' \cdot \mathbf{a}.$$

This has the same form as the Hamiltonian in the inertial frame, except for the additional term

$$V'(\mathbf{r}') = \sum_{i=1}^{N} m_i \mathbf{r}'_i \cdot \mathbf{a}$$

which acts like an additional potential. In the accelerating system there is thus an additional effective force

$$\mathbf{F}'_i = -\nabla'_i V' = -m_i \mathbf{a}$$

acting on the particles. This is as if there were a uniform gravitational field $-\mathbf{a}$ present in the accelerating frame.

CHAPTER VIII

HAMILTON-JACOBI THEORY

Exercise 8.01*

(a) Obtain Hamilton's principal function $S_H(z,t;z_0,t_0)$ for a particle of mass m which moves vertically in the uniform gravitational field g near the surface of the earth, by integrating the Lagrangian $L = \frac{1}{2}m\dot{z}^2 - mgz$ along the actual path which joins the end points. (Ans. $S_H = \frac{m}{2}\frac{(z-z_0)^2}{t-t_0} - \frac{1}{2}mg(z+z_0)(t-t_0) - \frac{1}{24}mg^2(t-t_0)^3$)

(b) Show that $S_H(z,t;z_0,t_0)$ is the type 1 generating function of a canonical transformation from present variables (z,p) to initial variables (z_0,p_0).

Solution

(a) Hamilton's principal function $S_H(z,t;z_0,t_0)$ for a particle of mass m which moves vertically in the uniform gravitational field g near the surface of the earth is given by

$$S_H(z,t;z_0,t_0) = \int_{t_0}^{t} L(z,\dot{z})\,dt'$$

where

$$L(z,\dot{z}) = \tfrac{1}{2}m\dot{z}^2 - mgz$$

is the Lagrangian and

$$z(t') = z_0 + v_0(t'-t_0) - \tfrac{1}{2}g(t'-t_0)^2 \qquad \dot{z}(t') = v(t') = v_0 - g(t'-t_0)$$

is the actual path. Here z_0 and v_0 are the initial (time t_0) position and velocity. The latter must be chosen so that the path passes through the final end point, position z at time t. This requires

$$v_0 = \frac{z-z_0}{t-t_0} + \frac{1}{2}g(t-t_0).$$

We have

$$S_H(z,t;z_0,t_0) = \int_{t_0}^{t}\left[\tfrac{1}{2}m\left(v_0 - g(t'-t_0)\right)^2 - mg\left(z_0 + v_0(t'-t_0) - \tfrac{1}{2}g(t'-t_0)^2\right)\right]dt'$$
$$= \left(\tfrac{1}{2}mv_0^2 - mgz_0\right)(t-t_0) - mgv_0(t-t_0)^2 + \tfrac{1}{3}mg^2(t-t_0)^3 .$$

Substituting for v_0, we find Hamilton's principal function

$$S_H(z,t;z_0,t_0) = \frac{m}{2}\frac{(z-z_0)^2}{t-t_0} - \frac{1}{2}mg(z+z_0)(t-t_0) - \frac{1}{24}mg^2(t-t_0)^3.$$

(b) To show that $S_H(z,t;z_0,t_0)$ is the type 1 generating function of a canonical transformation from present (time t) variables to initial (time t_0) variables, we evaluate its derivatives with respect to z and to z_0, obtaining

$$\frac{\partial S_H}{\partial z} = m\frac{z-z_0}{t-t_0} - \frac{1}{2}mg(t-t_0) = mv = p$$

where p is the present momentum and

$$\frac{\partial S_H}{\partial z_0} = -m\frac{z-z_0}{t-t_0} - \frac{1}{2}mg(t-t_0) = -mv_0 = -p_0$$

where p_0 is the initial momentum. These are as required.

Exercise 8.02

(a) Obtain a Jacobi complete integral $S_J(z,t;E) = W_J(z;E) - Et$ for a particle of mass m which moves vertically in the uniform gravitational field g near the surface of the earth, by integrating the time-independent Hamilton-Jacobi equation

$$\frac{1}{2m}\left(\frac{dW}{dz}\right)^2 + mgz = E.$$

(b) Use your solution to obtain the general solution $(z(t), p(t))$ to the dynamical problem.
(c) Obtain Hamilton's principal function from your Jacobi complete integral by setting $S_H(z,t;z_0,t_0) = S_J(z,t;E) - S_J(z_0,t_0;E)$ and then eliminating E by using $\partial S_H/\partial E = 0$.

Solution

(a) The time-independent Hamilton-Jacobi equation gives

$$W_J(z;E) = \int\sqrt{2m(E - mgz)}\,dz = -\frac{2}{3}\frac{\sqrt{2m}}{mg}(E - mgz)^{3/2},$$

so a Jacobi complete integral is

$$S_J(z,t;E) = -\frac{2}{3}\frac{\sqrt{2m}}{mg}(E - mgz)^{3/2} - Et.$$

(b) $S_J(z,t;E)$ is the type 2 generating function of a canonical transformation from old variables (z,p) to new variables (β_E, E) which are constant. Here β_E is the new "coordinate" conjugate to the new "momentum" E. The equations of the canonical transformation read

$$\beta_E = \frac{\partial S_J}{\partial E} = -\frac{\sqrt{2m}}{mg}(E - mgz)^{1/2} - t \quad \text{and} \quad p = \frac{\partial S_J}{\partial z} = \sqrt{2m(E - mgz)}.$$

The first of these gives the coordinate z as a function of time,

$$z(t) = \frac{E}{mg} - \frac{1}{2}g(t + \beta_E)^2,$$

and the second then gives the momentum p as a function of time,

$$p(t) = -mg(t + \beta_E).$$

These equations, which contain the two independent constants β_E and E, are the general solution to the dynamical problem. The constant $-\beta_E$ is the time at which the mass reaches its maximum height, and E/mg is the maximum height reached.

(c) Hamilton's principal function S_H for this system can be found from Jacobi's complete integral S_J by setting

$$\begin{aligned} S_H(z,t;z_0,t_0) &= S_J(z,t;E) - S_J(z_0,t_0;E) \\ &= -\frac{2}{3}\frac{\sqrt{2m}}{mg}\left[(E - mgz)^{3/2} - (E - mgz_0)^{3/2}\right] - E(t - t_0) \end{aligned}$$

and then expressing E in terms of $(z,t;z_0,t_0)$ by using the equation

$$\frac{\partial S_H}{\partial E} = -\frac{\sqrt{2m}}{mg}\left[(E - mgz)^{1/2} - (E - mgz_0)^{1/2}\right] - (t - t_0) = 0.$$

To solve this equation for E, we introduce auxiliary variables v_0 and v, setting

$$E = \tfrac{1}{2}mv_0^2 + mgz_0 \quad \text{and} \quad E = \tfrac{1}{2}mv^2 + mgz.$$

These equations here *define* the velocities v_0 and v. Together they imply that

$$v^2 - v_0^2 = -2g(z - z_0).$$

Further, the equation $\partial S_H/\partial E = 0$ becomes

$$v - v_0 = -g(t - t_0).$$

These last two equations give v_0 and v in terms of $(z, t; z_0, t_0)$,

$$v_0 = \frac{z - z_0}{t - t_0} + \frac{1}{2}g(t - t_0) \quad \text{and} \quad v = \frac{z - z_0}{t - t_0} - \frac{1}{2}g(t - t_0).$$

The energy is then

$$E = \frac{m}{2}\left(\frac{z - z_0}{t - t_0}\right)^2 + \frac{1}{2}mg(z + z_0) + \frac{1}{8}mg^2(t - t_0)^2,$$

and Hamilton's principal function becomes

$$\begin{aligned} S_H &= -\frac{m}{3g}(v^3 - v_0^3) - E(t - t_0) \\ &= \frac{m}{2}\frac{(z - z_0)^2}{t - t_0} - \frac{1}{2}mg(z + z_0)(t - t_0) - \frac{1}{24}mg^2(t - t_0)^3 . \end{aligned}$$

This agrees with the result obtained in Exercise 8.01.

Exercise 8.03*

(a) Obtain Hamilton's principal function $S_H(x,t;x_0,t_0)$ for the simple harmonic oscillator by integrating the Lagrangian $L = \frac{1}{2}m\dot{x}^2 - \frac{1}{2}m\omega^2 x^2$ along the actual path between the end points. (Ans. $S_H = \dfrac{m\omega}{2\sin\omega(t-t_0)}\left[\left(x^2+x_0^2\right)\cos\omega(t-t_0) - 2xx_0\right]$)

(b) Evaluate the action along the constant velocity path from (x_0,t_0) to (x,t), and compare with the result of (a). In particular, show for those paths which start at $(x_0 = 0, t_0 = 0)$ that S(actual path) < S(constant v path) provided $\omega t < \pi$.

(c) Show that $S_H(x,t;x_0,t_0)$ is the type 1 generating function of a canonical transformation from present variables (x,t) to initial variables (x_0,t_0).

Solution

(a) Hamilton's principal function $S_H(x,t;x_0,t_0)$ for a simple harmonic oscillator is given by

$$S_H(x,t;x_0,t_0) = \int_{t_0}^{t} L(x,\dot{x})\,dt'$$

where

$$L(x,\dot{x}) = \tfrac{1}{2}m\dot{x}^2 - \tfrac{1}{2}m\omega^2 x^2$$

is the Lagrangian and

$$\begin{aligned} x(t') &= x_0\cos\omega(t'-t_0) + (v_0/\omega)\sin\omega(t'-t_0) \\ \dot{x}(t') &= -\omega x_0\sin\omega(t'-t_0) + v_0\cos\omega(t'-t_0), \end{aligned}$$

is the actual path. Here x_0 and v_0 are the initial position and velocity. The latter must be chosen so that the path passes through the final end point. This requires

$$v_0 = \omega\frac{x - x_0\cos\omega(t-t_0)}{\sin\omega(t-t_0)}.$$

We have

$$S_H = \int_{t_0}^{t}\Big[\tfrac{1}{2}m\big(-\omega x_0 \sin\omega(t'-t_0) + v_0\cos\omega(t'-t_0)\big)^2$$
$$-\tfrac{1}{2}m\omega^2\big(x_0\cos\omega(t'-t_0) + (v_0/\omega)\sin\omega(t'-t_0)\big)^2\Big]dt'$$
$$= \int_{t_0}^{t}\Big[\Big(\tfrac{1}{2}mv_0^2 - \tfrac{1}{2}m\omega^2 x_0^2\Big)\cos 2\omega(t'-t_0) - m\omega x_0 v_0 \sin 2\omega(t'-t_0)\Big]dt'$$
$$= \Big(\tfrac{1}{2}mv_0^2 - \tfrac{1}{2}m\omega^2x_0^2\Big)\big(\sin 2\omega(t-t_0)/2\omega\big) - m\omega x_0 v_0\big(\big(1-\cos 2\omega(t-t_0)\big)/2\omega\big).$$

Substituting for v_0, we find Hamilton's principal function

$$S_H = \frac{m\omega\cos\omega(t-t_0)}{2\sin\omega(t-t_0)}\Big[\big(x - x_0\cos\omega(t-t_0)\big)^2 - x_0^2\sin^2\omega(t-t_0)\Big]$$
$$-m\omega x_0\big(x - x_0\cos\omega(t-t_0)\big)\sin\omega(t-t_0)$$
$$= \frac{m\omega}{2\sin\omega(t-t_0)}\Big[\big(x^2 + x_0^2\big)\cos\omega(t-t_0) - 2xx_0\Big].$$

(b) We now wish to find the action for the simple harmonic oscillator along the constant velocity path

$$x(t') = x_0 + v_0(t'-t_0), \qquad \dot{x}(t') = v_0,$$

between the end points (x_0, t_0) and (x, t). Here x_0 is the initial position and v_0 is the initial velocity. The latter must be chosen so that the path passes through the final end point. This requires

$$v_0 = \frac{x - x_0}{t - t_0}.$$

The action is then

$$S = \int_{t_0}^{t}\Big[\tfrac{1}{2}mv_0^2 - \tfrac{1}{2}m\omega^2\big(x_0 + v_0(t'-t_0)\big)^2\Big]dt'$$
$$= \tfrac{1}{2}mv_0^2(t'-t_0) - \tfrac{1}{6}(m\omega^2/v_0)\Big(\big(x_0 + v_0(t-t_0)\big)^3 - x_0^3\Big).$$

Substituting for v_0, we find

$$S = \frac{m}{2}\frac{(x-x_0)^2}{t-t_0} - \frac{1}{6}m\omega^2(x^2 + xx_0 + x_0^2)(t-t_0).$$

Let us compare the actions along the actual path and along the constant velocity path for those paths which start at $x_0 = 0$ at $t_0 = 0$. For these, the action along the actual path (Hamilton's principal function) is

$$S(\text{actual path}) = \frac{mx^2}{2t}(\omega t \cot \omega t),$$

and the action along the constant velocity path is

$$S(\text{constant v path}) = \frac{mx^2}{2t}\left(1 - \tfrac{1}{3}(\omega t)^2\right).$$

The factors in brackets in these expressions are shown in Fig. 1, from which it is clear that

$$S(\text{actual path}) < S(\text{constant v path})$$

provided $\omega t < \pi$. In particular, for small ωt we can expand

$$\omega t \cot \omega t = 1 - \tfrac{1}{3}(\omega t)^2 - \tfrac{1}{45}(\omega t)^4 - \cdots < 1 - \tfrac{1}{3}(\omega t)^2.$$

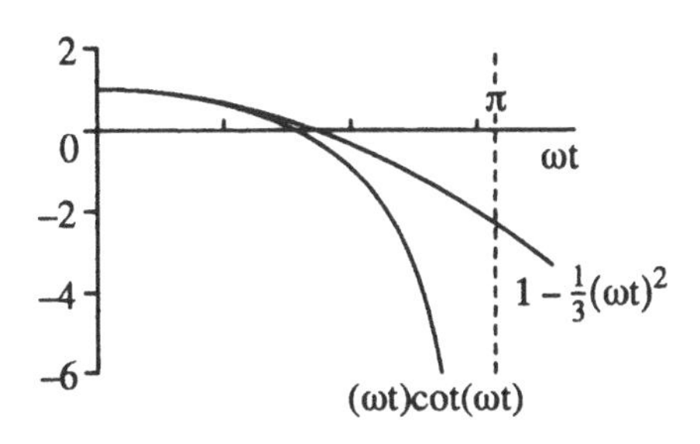

Ex. 8.03, Fig. 1

(c) To show that $S_H(x,t;x_0,t_0)$ is the type 1 generating function of a canonical transformation from present (time t) variables to initial (time t_0) variables, we evaluate its derivatives with respect to x and to x_0, obtaining

$$\frac{\partial S_H}{\partial x} = \frac{m\omega}{\sin \omega(t - t_0)}\left(x \cos \omega(t - t_0) - x_0\right) = p$$

where p is the present momentum and

$$\frac{\partial S_H}{\partial x_0} = \frac{m\omega}{\sin \omega(t - t_0)}\left(x_0 \cos \omega(t - t_0) - x\right) = -p_0$$

where p_0 is the initial momentum. These are as required.

Exercise 8.04*

The motion of a projectile near the surface of the earth is governed by the Hamiltonian

$$H = \frac{1}{2m}\left(p_x^2 + p_y^2\right) + mgy$$

where x denotes the horizontal coordinate and y the vertical, and p_x and p_y are their conjugate momenta.
(a) Set up and find a complete integral W to the time-independent Hamilton-Jacobi equation.
(b) Use your solution to obtain x and y as functions of t.

Solution

(a) The time-independent Hamilton-Jacobi equation is

$$\frac{1}{2m}\left(\left(\frac{\partial W}{\partial x}\right)^2 + \left(\frac{\partial W}{\partial y}\right)^2\right) + mgy = E.$$

We try a solution of the form

$$W = X(x) + Y(y).$$

The Hamilton-Jacobi equation becomes

$$\frac{1}{2m}\left(\frac{dX}{dx}\right)^2 + \frac{1}{2m}\left(\frac{dY}{dy}\right)^2 + mgy = E.$$

The first term on the left is a function only of x and the remaining terms are functions only of y. The variables are completely separated, and we can set

$$\frac{1}{2m}\left(\frac{dX}{dx}\right)^2 = \frac{\alpha^2}{2m}, \qquad \frac{1}{2m}\left(\frac{dY}{dy}\right)^2 + mgy = E - \frac{\alpha^2}{2m},$$

where α is a separation constant. These equations give

$$X(x) = \alpha x, \qquad Y(y) = \int\sqrt{2m(E - \alpha^2/2m - mgy)}\,dy = -\frac{2}{3}\frac{\sqrt{2m}}{mg}(E - \alpha^2/2m - mgy)^{3/2},$$

so a complete integral to the time-independent Hamilton-Jacobi equation is

$$W(x,y;\alpha,E) = \alpha x - \frac{2}{3}\frac{\sqrt{2m}}{mg}(E - \alpha^2/2m - mgy)^{3/2}.$$

(b) The complete integral W is the generating function of the canonical transformation to new variables which are cyclic. We have

$$\beta = \frac{\partial W}{\partial \alpha} = x + \frac{\alpha}{m^2 g}\sqrt{2m(E - \alpha^2/2m - mgy)}, \qquad p_x = \frac{\partial W}{\partial x} = \alpha,$$

$$\beta_E + t = \frac{\partial W}{\partial E} = -\frac{1}{mg}\sqrt{2m(E - \alpha^2/2m - mgy)}, \quad p_y = \frac{\partial W}{\partial y} = \sqrt{2m(E - \alpha^2/2m - mgy)}.$$

These equations give the general solution to the projectile problem,

$$x(t) = \beta + (\alpha/m)(t + \beta_E), \qquad p_x(t) = \alpha,$$

$$y(t) = \frac{(E - \alpha^2/2m)}{mg} - \frac{1}{2}g(t + \beta_E)^2, \qquad p_y(t) = -mg(t + \beta_E).$$

We can also identify the three time-independent constants of the motion

$$\alpha = p_x, \quad mg\beta = mgx + p_x p_y/m, \quad \text{and} \quad E = (p_x^2 + p_y^2)/2m + mgy,$$

associated with this system.

Exercise 8.05

The motion of a free particle on a plane can be described by the Hamiltonian

$$H = \frac{1}{2m}\left(p_r^2 + \frac{p_\phi^2}{r^2}\right)$$

where p_r and p_ϕ are the momenta conjugate to the plane polar coordinates r and ϕ.
(a) Set up and find a complete integral W to the time-independent Hamilton-Jacobi equation.
(b) Use your solution to obtain r and ϕ as functions of t.

Solution

(a) The time-independent Hamilton-Jacobi equation is

$$\frac{1}{2m}\left(\left(\frac{\partial W}{\partial r}\right)^2 + \frac{1}{r^2}\left(\frac{\partial W}{\partial \phi}\right)^2\right) = E.$$

We try a solution of the form

$$W = R(r) + \Phi(\phi).$$

The Hamilton-Jacobi equation becomes

$$\frac{1}{2m}\left(\left(\frac{dR}{dr}\right)^2 + \frac{1}{r^2}\left(\frac{d\Phi}{d\phi}\right)^2\right) = E.$$

This can be rearranged in the form

$$2mr^2\left(\frac{1}{2m}\left(\frac{dR}{dr}\right)^2 - E\right) = -\left(\frac{d\Phi}{d\phi}\right)^2.$$

The left-hand side is a function only of r and the right-hand side is a function only of ϕ. Both sides must equal a (negative) constant $-L^2$. The constant L turns out to be the angular momentum. We have

$$\frac{1}{2m}\left(\frac{dR}{dr}\right)^2 + \frac{L^2}{2mr^2} = E \quad \text{and} \quad \left(\frac{d\Phi}{d\phi}\right)^2 = L^2.$$

These yield

$$R = \int\sqrt{2mE - L^2/r^2}\,dr \quad \text{and} \quad \Phi = L\phi,$$

so a complete integral to the time-independent Hamilton-Jacobi equation is

$$W = \int\sqrt{2mE - L^2/r^2}\,dr + L\phi.$$

It is a function of r and ϕ and contains the two non-additive constants E and L.

(b) The Jacobi complete integral $W(r,\phi;E,L)$ is the generating function of a canonical transformation to new canonical variables, where the new momenta are E and L and the new coordinates are cyclic. The equations of the canonical transformation read

$$p_r = \frac{\partial W}{\partial r} = \sqrt{2mE - L^2/r^2}, \qquad p_\phi = \frac{\partial W}{\partial \phi} = L,$$

$$\beta_E + t = \frac{\partial W}{\partial E} = \int\frac{m\,dr}{\sqrt{2mE - L^2/r^2}}, \quad \beta_L = \frac{\partial W}{\partial L} = \int\frac{-L}{\sqrt{2mE - L^2/r^2}}\frac{dr}{r^2} + \phi.$$

The r-integration in the "β_L-equation" can be performed by setting

$$\frac{1}{r}=\frac{\sqrt{2mE}}{L}\cos\alpha, \qquad \frac{dr}{r^2}=\frac{\sqrt{2mE}}{L}\sin\alpha\, d\alpha,$$

and gives $\beta_L=-\alpha+\phi$. We thus find the equation of the trajectory

$$r\cos(\phi-\beta_L)=\frac{L}{\sqrt{2mE}}.$$

This is the equation of a straight line in polar coordinates; see Fig. 1.

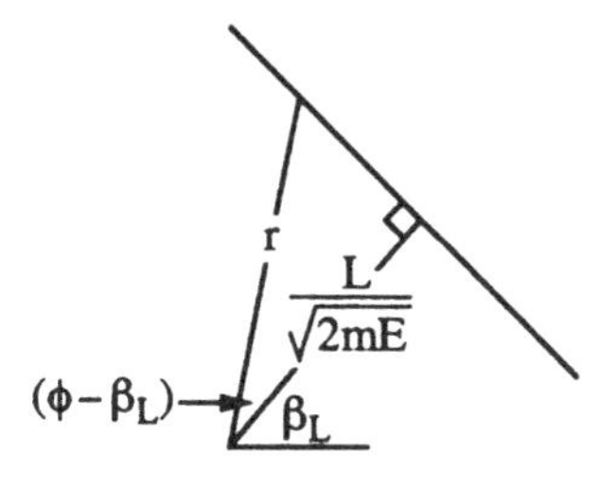

Ex. 8.05, Fig. 1

The r-integration in the "β_E-equation" gives

$$\beta_E+t=\sqrt{\frac{m}{2E}}\int\frac{r\,dr}{\sqrt{r^2-L^2/2mE}}=\sqrt{\frac{m}{2E}}\sqrt{r^2-L^2/2mE},$$

which can be rearranged as

$$r^2=\frac{L^2}{2mE}+\frac{2E}{m}(t+\beta_E)^2.$$

This says that the particle moves along the trajectory at constant speed $v=\sqrt{2E/m}$. It is convenient to let $b=L/\sqrt{2mE}$ denote the "impact parameter" or distance of closest approach to the origin. The preceding equations of the trajectory can then be written

$$r\cos(\phi-\beta_L)=b \quad \text{and} \quad r^2=b^2+v^2(t+\beta_E)^2.$$

Exercise 8.06

Use the Hamilton-Jacobi method to find the general equations of motion for a three-dimensional isotropic harmonic oscillator with potential

$$V = \tfrac{1}{2}k(x^2 + y^2 + z^2) = \tfrac{1}{2}kr^2.$$

(a) First use cartesian coordinates (x, y, z).
(b) Then do the problem again using spherical polar coordinates (r, θ, ϕ).

Solution

(a) The Hamiltonian for a three-dimensional isotropic harmonic oscillator is, in cartesian coordinates,

$$H = \frac{1}{2m}\left(p_x^2 + p_y^2 + p_z^2\right) + \frac{1}{2}k\left(x^2 + y^2 + z^2\right).$$

The time-independent Hamilton-Jacobi equation is

$$\frac{1}{2m}\left(\left(\frac{\partial W}{\partial x}\right)^2 + \left(\frac{\partial W}{\partial y}\right)^2 + \left(\frac{\partial W}{\partial z}\right)^2\right) + \frac{1}{2}k\left(x^2 + y^2 + z^2\right) = E.$$

We try a solution of the form

$$W = X(x) + Y(y) + Z(z).$$

The Hamilton-Jacobi equation becomes

$$\left[\frac{1}{2m}\left(\frac{dX}{dx}\right)^2 + \frac{1}{2}kx^2\right] + \left[\frac{1}{2m}\left(\frac{dY}{dy}\right)^2 + \frac{1}{2}ky^2\right] + \left[\frac{1}{2m}\left(\frac{dZ}{dz}\right)^2 + \frac{1}{2}kz^2\right] = E.$$

The first term on the left is a function only of x, the second is a function only of y, and the third is a function only of z. Since the sum of the three terms is a constant, each must equal a constant. We set

$$\frac{1}{2m}\left(\frac{dX}{dx}\right)^2 + \frac{1}{2}kx^2 = \alpha_x, \quad \frac{1}{2m}\left(\frac{dY}{dy}\right)^2 + \frac{1}{2}ky^2 = \alpha_y, \quad \frac{1}{2m}\left(\frac{dZ}{dz}\right)^2 + \frac{1}{2}kz^2 = \alpha_z,$$

where α_x, α_y, and α_z are separation constants with $\alpha_x + \alpha_y + \alpha_z = E$. These give

$$X = \int\sqrt{2m\alpha_x - mkx^2}\,dx, \quad Y = \int\sqrt{2m\alpha_y - mky^2}\,dy, \quad Z = \int\sqrt{2m\alpha_z - mkz^2}\,dz,$$

so a complete integral to the time-independent Hamilton-Jacobi equation is

$$W = \int\sqrt{2m\alpha_x - mkx^2}\,dx + \int\sqrt{2m\alpha_y - mky^2}\,dy + \int\sqrt{2m\alpha_z - mkz^2}\,dz.$$

The canonical transformation generated by W gives

$$\beta_x + t = \frac{\partial W}{\partial \alpha_x} = m\int\frac{dx}{\sqrt{2m\alpha_x - mkx^2}}$$

plus two similar equations for the y and z coordinates. The integration over x can be performed by setting

$$x = \sqrt{\frac{2\alpha_x}{k}}\sin\phi_x, \qquad dx = \sqrt{\frac{2\alpha_x}{k}}\cos\phi_x\,d\phi_x.$$

We find

$$\phi_x = \sqrt{\frac{k}{m}}(t+\beta_x), \quad \text{so} \quad x(t) = \sqrt{\frac{2\alpha_x}{k}}\sin\sqrt{\frac{k}{m}}(t+\beta_x).$$

Similarly, we find

$$y(t) = \sqrt{\frac{2\alpha_y}{k}}\sin\sqrt{\frac{k}{m}}(t+\beta_y) \quad \text{and} \quad z(t) = \sqrt{\frac{2\alpha_z}{k}}\sin\sqrt{\frac{k}{m}}(t+\beta_z).$$

(b) The Hamiltonian for a three-dimensional isotropic harmonic oscillator is, in spherical polar coordinates,

$$H = \frac{1}{2m}\left(p_r^2 + \frac{1}{r^2}p_\theta^2 + \frac{1}{r^2\sin^2\theta}p_\phi^2\right) + \frac{1}{2}kr^2.$$

The time-independent Hamilton-Jacobi equation is

$$\frac{1}{2m}\left(\left(\frac{\partial W}{\partial r}\right)^2 + \frac{1}{r^2}\left(\frac{\partial W}{\partial \theta}\right)^2 + \frac{1}{r^2\sin^2\theta}\left(\frac{\partial W}{\partial \phi}\right)^2\right) + \frac{1}{2}kr^2 = E.$$

Separation of variables proceeds as in *Lagrangian and Hamiltonian Mechanics* (pages 157, 158), and we find a complete integral

$$W = \sqrt{2m}\int\sqrt{E - \frac{1}{2}kr^2 - \frac{L^2}{2mr^2}}\,dr + \int\sqrt{L^2 - \frac{L_z^2}{\sin^2\theta}}\,d\theta + L_z\phi.$$

The canonical transformation generated by W gives

$$\beta_E + t = \frac{\partial W}{\partial E} = \sqrt{\frac{m}{2}} \int \frac{dr}{\sqrt{E - \frac{1}{2}kr^2 - \frac{L^2}{2mr^2}}},$$

$$\beta_L = \frac{\partial W}{\partial L} = \sqrt{2m} \int \frac{dr}{\sqrt{E - \frac{1}{2}kr^2 - \frac{L^2}{2mr^2}}} \left(\frac{-L}{2mr^2}\right) + \int \frac{L\,d\theta}{\sqrt{L^2 - \frac{L_z^2}{\sin^2\theta}}},$$

$$\beta_{L_z} = \frac{\partial W}{\partial L_z} = \int \frac{d\theta}{\sqrt{L^2 - \frac{L_z^2}{\sin^2\theta}}} \left(\frac{-L_z}{\sin^2\theta}\right) + \phi\,.$$

The integration of the third of these equations is done in *Lagrangian and Hamiltonian Mechanics* (pages 159, 160) with the result

$$\sin i \sin\theta \sin(\phi - \beta_{L_z}) = \cos i \cos\theta.$$

This says that the orbit lies in a plane through the origin. The θ-integration in the second equation is also done in *Lagrangian and Hamiltonian Mechanics* (pages 161, 162), so this equation becomes

$$\bar{\theta} - \beta_L = \frac{L}{\sqrt{2m}} \int \frac{dr}{r^2\sqrt{E - \frac{1}{2}kr^2 - \frac{L^2}{2mr^2}}}.$$

Here $\bar{\theta}$ is the angle, measured in the plane of the orbit, between the radius vector and a reference direction. The r-integration here is done in Exercise 1.14 with the result

$$\frac{1}{r^2} = \frac{mE}{L^2}\left[1 + \sqrt{1 - \frac{kL^2}{mE^2}} \cos 2(\bar{\theta} - \beta_L)\right].$$

This says that the orbit is an ellipse with geometric center at the force center. The r-integration in the first of the canonical transformation equations is also done in Exercise 1.14 with the result

$$r^2 = \frac{E}{k}\left[1 - \sqrt{1 - \frac{kL^2}{mE^2}} \cos 2\sqrt{\frac{k}{m}}(t + \beta_E)\right].$$

This says how the particle moves along the orbit in time.

Exercise 8.07

Use the Hamilton-Jacobi method to study the motion of a particle in a dipole field with (non-central) potential

$$V = \frac{k\cos\theta}{r^2}.$$

(a) Write down the time-independent Hamilton-Jacobi equation for W in spherical polar coordinates.
(b) Show that this equation can be solved by the method of separation of variables, and obtain an expression for W of the form $W = W(r,\theta,\phi;E,\alpha_2,\alpha_3)$. Your answer will also involve certain integrals; you need not evaluate these at this stage.
(c) Interpret physically your separation constants α_2,α_3 by obtaining p_r, p_θ, p_ϕ in terms of $r,\theta,\phi,E,\alpha_2,\alpha_3$. Hence show that the z-component L_z of the angular momentum of the particle is constant, and further that $L^2 + 2mk\cos\theta$ is constant, where L^2 is the square of the total angular momentum of the particle.
(d) By considering the equation

$$\frac{\partial W}{\partial E} = t + \beta_1$$

find how r varies with time.

Solution

(a) The Hamiltonian for a particle moving in a dipole field is, in spherical polar coordinates,

$$H = \frac{1}{2m}\left(p_r^2 + \frac{1}{r^2}p_\theta^2 + \frac{1}{r^2\sin^2\theta}p_\phi^2\right) + \frac{k\cos\theta}{r^2},$$

so the time-independent Hamilton-Jacobi equation is

$$\frac{1}{2m}\left(\left(\frac{\partial W}{\partial r}\right)^2 + \frac{1}{r^2}\left(\frac{\partial W}{\partial \theta}\right)^2 + \frac{1}{r^2\sin^2\theta}\left(\frac{\partial W}{\partial \phi}\right)^2\right) + \frac{k\cos\theta}{r^2} = E.$$

(b) We try a solution of the form

$$W = R(r) + \Theta(\theta) + \Phi(\phi).$$

The Hamilton-Jacobi equation becomes

$$\frac{1}{2m}\left(\left(\frac{dR}{dr}\right)^2 + \frac{1}{r^2}\left(\frac{d\Theta}{d\theta}\right)^2 + \frac{1}{r^2\sin^2\theta}\left(\frac{d\Phi}{d\phi}\right)^2\right) + \frac{k\cos\theta}{r^2} = E.$$

This can be rearranged in the form

$$2mr^2\sin^2\theta\left(\frac{1}{2m}\left(\frac{dR}{dr}\right)^2 + \frac{1}{2mr^2}\left(\frac{d\Theta}{d\theta}\right)^2 + \frac{k\cos\theta}{r^2} - E\right) = -\left(\frac{d\Phi}{d\phi}\right)^2.$$

The left-hand side is a function only of r and θ, and the right-hand side is a function only of ϕ. Both sides must equal a (negative) constant. We set

$$2mr^2\sin^2\theta\left(\frac{1}{2m}\left(\frac{dR}{dr}\right)^2 + \frac{1}{2mr^2}\left(\frac{d\Theta}{d\theta}\right)^2 + \frac{k\cos\theta}{r^2} - E\right) = -L_z^2, \qquad \left(\frac{d\Phi}{d\phi}\right)^2 = L_z^2.$$

The separation constant L_z turns out to be the z-component of the angular momentum. The rθ-equation can be rearranged in the form

$$2mr^2\left(\frac{1}{2m}\left(\frac{dR}{dr}\right)^2 - E\right) = -\left(\left(\frac{d\Theta}{d\theta}\right)^2 + \frac{L_z^2}{\sin^2\theta} + 2mk\cos\theta\right).$$

The left-hand side is a function only of r, and the right-hand side is a function only of θ. Both sides must equal a constant. We set

$$2mr^2\left(\frac{1}{2m}\left(\frac{dR}{dr}\right)^2 - E\right) = -\alpha, \qquad \left(\frac{d\Theta}{d\theta}\right)^2 + \frac{L_z^2}{\sin^2\theta} + 2mk\cos\theta = \alpha.$$

Note that the separation constant α is not necessarily positive. The variables are now completely separated, and we can integrate to obtain

$$R = \int\sqrt{2mE - \frac{\alpha}{r^2}}\,dr, \qquad \Theta = \int\sqrt{\alpha - \frac{L_z^2}{\sin^2\theta} - 2mk\cos\theta}\,d\theta, \qquad \Phi = L_z\phi.$$

A complete integral to the Hamilton-Jacobi equation is thus

$$W = \int\sqrt{2mE - \frac{\alpha}{r^2}}\,dr + \int\sqrt{\alpha - \frac{L_z^2}{\sin^2\theta} - 2mk\cos\theta}\,d\theta + L_z\phi.$$

(c) The complete integral W is the type 2 generating function of a canonical transformation. Half of the transformation equations reads

$$p_r = \frac{\partial W}{\partial r} = \sqrt{2mE - \frac{\alpha}{r^2}}, \quad p_\theta = \frac{\partial W}{\partial \theta} = \sqrt{\alpha - \frac{L_z^2}{\sin^2\theta} - 2mk\cos\theta}, \quad p_\phi = \frac{\partial W}{\partial \phi} = L_z.$$

Since p_ϕ is the z-component of the angular momentum (see *Lagrangian and Hamiltonian Mechanics*, page 158), this identifies the first separation constant L_z. Further, since $p_\theta^2 + p_\phi^2/\sin^2\theta$ is the square of the total angular momentum, L^2, the second separation constant α is

$$\alpha = L^2 + 2mk\cos\theta.$$

(d) The other half of the transformation equations reads

$$\beta_E + t = \frac{\partial W}{\partial E} = m\int \frac{dr}{\sqrt{2mE - \alpha/r^2}},$$

$$\beta_\alpha = \frac{\partial W}{\partial \alpha} = \frac{1}{2}\int \frac{(-dr/r^2)}{\sqrt{2mE - \alpha/r^2}} + \frac{1}{2}\int \frac{d\theta}{\sqrt{\alpha - L_z^2/\sin^2\theta - 2mk\cos\theta}},$$

$$\beta_{L_z} = \frac{\partial W}{\partial L_z} = \int \frac{d\theta}{\sqrt{\alpha - L_z^2/\sin^2\theta - 2mk\cos\theta}}\left(\frac{-L_z}{\sin^2\theta}\right) + \phi.$$

The r-integration in the first of these is easily performed,

$$\beta_E + t = m\int \frac{r\,dr}{\sqrt{2mEr^2 - \alpha}} = \frac{1}{2E}\sqrt{2mEr^2 - \alpha},$$

and the result can be rearranged to give

$$r^2 = \frac{\alpha}{2mE} + \frac{2E}{m}(t + \beta_E)^2.$$

The nature of the radial motion depends on the signs of E and α. If both are positive, the particle comes in from infinity, reaches a minimum radius of $\sqrt{\alpha/2mE}$ at $t + \beta_E = 0$, and goes out again to infinity. If E is positive but α negative, the particle, starting at $t = -\infty$, comes in from infinity and reaches the origin $r = 0$ at $t + \beta_E = -\sqrt{|\alpha|}/2E$ (or it starts at the origin $r = 0$ at $t + \beta_E = \sqrt{|\alpha|}/2E$ and goes out to infinity $t = \infty$). If both E and α are negative, the particle is confined to the region $r < \sqrt{\alpha/2mE}$. This maximum radius occurs for $t + \beta_E = 0$, and the particle reaches $r = 0$ at $t + \beta_E = \pm\sqrt{|\alpha|}/2E$.

Exercise 8.08

A particle of mass m moves in a field which is a superposition of a Coulomb field with potential $-k/r$ and a constant field F in the z-direction with potential $-Fz$. The total potential is

$$V = -\frac{k}{r} - Fz.$$

(a) Set up the time-independent Hamilton-Jacobi equation in paraboloidal coordinates (see Exercises 3.09 and 6.04). Show that the variables separate, and obtain an expression for the Jacobi function W of the form

$$W = \sqrt{2m}\int\sqrt{k - \alpha/m - L_z^2/2m\xi^2 + F\xi^4/2 + E\xi^2}\,d\xi$$
$$+\sqrt{2m}\int\sqrt{k + \alpha/m - L_z^2/2m\eta^2 - F\eta^4/2 + E\eta^2}\,d\eta + L_z\phi$$

where L_z and α are separation constants.
(b) Interpret physically the separation constants, showing that L_z is the z-component of the angular momentum, and that $\alpha = K_z + \frac{1}{2}mF(r^2 - z^2)$ where K_z is the z-component of the Laplace-Runge-Lenz vector (see Exercise 1.12).

Solution

(a) The Hamiltonian for a particle which moves in a superposition of a Coulomb field and a constant field is, in paraboloidal coordinates,

$$H = \frac{1}{2m(\xi^2 + \eta^2)}(p_\xi^2 + p_\eta^2) + \frac{1}{2m\xi^2\eta^2}p_\phi^2 - \frac{2k}{\xi^2 + \eta^2} - \frac{1}{2}F(\xi^2 - \eta^2).$$

The kinetic energy in this expression comes from Exercise 6.04 which in turn is built on Exercise 3.09. The potential energy is written in terms of paraboloidal coordinates by setting $r = \frac{1}{2}(\xi^2 + \eta^2)$ and $z = \frac{1}{2}(\xi^2 - \eta^2)$. The time-independent Hamilton-Jacobi equation is

$$\frac{1}{2m(\xi^2 + \eta^2)}\left(\left(\frac{\partial W}{\partial \xi}\right)^2 + \left(\frac{\partial W}{\partial \eta}\right)^2\right) + \frac{1}{2m\xi^2\eta^2}\left(\frac{\partial W}{\partial \phi}\right)^2 - \frac{2k}{\xi^2 + \eta^2} - \frac{1}{2}F(\xi^2 - \eta^2) = E.$$

We try a solution of the form

$$W = X(\xi) + Y(\eta) + \Phi(\phi).$$

The Hamilton-Jacobi equation becomes

$$\frac{1}{2m(\xi^2+\eta^2)}\left(\left(\frac{dX}{d\xi}\right)^2+\left(\frac{dY}{d\eta}\right)^2\right)+\frac{1}{2m\xi^2\eta^2}\left(\frac{d\Phi}{d\phi}\right)^2-\frac{2k}{\xi^2+\eta^2}-\frac{1}{2}F(\xi^2-\eta^2)=E.$$

This can be rearranged in the form

$$2m\xi^2\eta^2\left[\frac{1}{2m(\xi^2+\eta^2)}\left(\left(\frac{dX}{d\xi}\right)^2+\left(\frac{dY}{d\eta}\right)^2\right)-\frac{2k}{\xi^2+\eta^2}-\frac{1}{2}F(\xi^2-\eta^2)-E\right]=-\left(\frac{d\Phi}{d\phi}\right)^2.$$

The left-hand side is a function only of ξ and η, and the right-hand side is a function only of ϕ. Both sides must equal a (negative) constant. We set

$$2m\xi^2\eta^2\left[\frac{1}{2m(\xi^2+\eta^2)}\left(\left(\frac{dX}{d\xi}\right)^2+\left(\frac{dY}{d\eta}\right)^2\right)-\frac{2k}{\xi^2+\eta^2}-\frac{1}{2}F(\xi^2-\eta^2)-E\right]=-L_z^2,$$

$$\left(\frac{d\Phi}{d\phi}\right)^2=L_z^2.$$

The $\xi\eta$-equation can be rearranged as

$$\left[\frac{1}{2m}\left(\frac{dX}{d\xi}\right)^2+\frac{L_z^2}{2m\xi^2}-\frac{1}{2}F\xi^4-E\xi^2\right]+\left[\frac{1}{2m}\left(\frac{dY}{d\eta}\right)^2+\frac{L_z^2}{2m\eta^2}+\frac{1}{2}F\eta^4-E\eta^2\right]=2k.$$

The first term on the left is a function only of ξ and the second is a function only of η. Since their sum is a constant, each must equal a constant. We set

$$\frac{1}{2m}\left(\frac{dX}{d\xi}\right)^2+\frac{L_z^2}{2m\xi^2}-\frac{1}{2}F\xi^4-E\xi^2=k-\frac{\alpha}{m},$$

$$\frac{1}{2m}\left(\frac{dY}{d\eta}\right)^2+\frac{L_z^2}{2m\eta^2}+\frac{1}{2}F\eta^4-E\eta^2=k+\frac{\alpha}{m},$$

where this choice of separation constant α turns out to be convenient. The variables are now completely separated, and we can integrate to obtain X, Y, and Φ. Their sum gives the complete integral

$$W=\sqrt{2m}\int\sqrt{k-\alpha/m-L_z^2/2m\xi^2+F\xi^4/2+E\xi^2}\,d\xi$$
$$+\sqrt{2m}\int\sqrt{k+\alpha/m-L_z^2/2m\eta^2-F\eta^4/2+E\eta^2}\,d\eta+L_z\phi.$$

(b) The momenta conjugate to (ξ, η, ϕ) are

$$p_\xi = \frac{\partial W}{\partial \xi} = \sqrt{2m}\sqrt{k - \frac{\alpha}{m} - \frac{L_z^2}{2m\xi^2} + \frac{1}{2}F\xi^4 + E\xi^2},$$

$$p_\eta = \frac{\partial W}{\partial \eta} = \sqrt{2m}\sqrt{k + \frac{\alpha}{m} - \frac{L_z^2}{2m\eta^2} - \frac{1}{2}F\eta^4 + E\eta^2},$$

$$p_\phi = \frac{\partial W}{\partial \phi} = L_z.$$

The third of these equations shows that the separation constant L_z is the z-component of the angular momentum. To identify the separation constant α, we begin by squaring and subtracting the first two equations, obtaining

$$4\alpha = p_\eta^2 - p_\xi^2 + L_z^2(1/\eta^2 - 1/\xi^2) + mF(\xi^4 + \eta^4) + 2mE(\xi^2 - \eta^2).$$

Substituting for $L_z = p_\phi$ and for

$$E = \frac{1}{2m(\xi^2 + \eta^2)}(p_\xi^2 + p_\eta^2) + \frac{1}{2m\xi^2\eta^2}p_\phi^2 - \frac{2k}{\xi^2 + \eta^2} - \frac{1}{2}F(\xi^2 - \eta^2),$$

we find

$$2\alpha = \frac{1}{\xi^2 + \eta^2}(\xi^2 p_\eta^2 - \eta^2 p_\xi^2) + \frac{\xi^2 - \eta^2}{\xi^2\eta^2}p_\phi^2 - 2mk\frac{\xi^2 - \eta^2}{\xi^2 + \eta^2} + mF\xi^2\eta^2.$$

This expresses α in terms of the paraboloidal canonical variables. Since, however, these variables are not in common usage, the expression may not be terribly illuminating. The symmetry of the system suggests we try rewriting α in terms of cylindrical coordinates

$$\rho\ (= \sqrt{x^2 + y^2}) = \xi\eta \quad \text{and} \quad z = \tfrac{1}{2}(\xi^2 - \eta^2)$$

(the spherical polar coordinate is $r = \sqrt{x^2 + y^2 + z^2} = \sqrt{\rho^2 + z^2} = \frac{1}{2}(\xi^2 + \eta^2)$). The transformation to these coordinates is a canonical transformation generated by $F_2 = \xi\eta p_\rho + \frac{1}{2}(\xi^2 - \eta^2)p_z$. Besides the above relations, we have the relations between the momenta

$$p_\xi = \frac{\partial F_2}{\partial \xi} = \eta p_\rho + \xi p_z \quad \text{and} \quad p_\eta = \frac{\partial F_2}{\partial \eta} = \xi p_\rho - \eta p_z.$$

The separation constant α becomes

$$\alpha = p_\rho(zp_\rho - \rho p_z) + zp_\phi^2/\rho^2 - mk\,z/r + \tfrac{1}{2}mF\rho^2 .$$

Now p_ρ and p_z are the ρ- and z-components of the linear momentum, so the term in brackets on the right is the ϕ-component of the angular momentum; it is L_ϕ. As we have seen, p_ϕ is the z-component of the angular momentum; it is L_z. The first two terms on the right are thus $p_\rho L_\phi - (L_z/\rho)(-zL_z/\rho)$. Now (L_z/ρ) is the ϕ-component of the linear momentum and $(-zL_z/\rho)$ is the ρ-component of the angular momentum, so the first two terms are the z-component of $\mathbf{p}\times\mathbf{L}$. The third term in α is the z-component of $-mk\hat{\mathbf{r}}$, so we finally obtain

$$\alpha = \left(\mathbf{p}\times\mathbf{L} - mk\hat{\mathbf{r}}\right)_z + \tfrac{1}{2}mF\rho^2 = K_z + \tfrac{1}{2}mF\rho^2$$

where K_z is the z-component of the Laplace-Runge-Lenz vector.

Exercise 8.09

(a) Write down the Hamilton-Jacobi equation for a particle of mass m and charge e in an electromagnetic field described by a scalar potential ϕ and a vector potential $\mathbf{A}$.
(b) Show that the Hamilton-Jacobi equation is invariant under a gauge transformation,

$$\phi' = \phi - (1/c)\,\partial\lambda/\partial t, \qquad \mathbf{A}' = \mathbf{A} + \nabla\lambda,$$

provided the Hamilton-Jacobi function is also transformed,

$$S' = S + (e/c)\lambda.$$

Solution

(a) The Hamiltonian for a particle of mass m and charge e in an electromagnetic field which is described by a scalar potential ϕ and a vector potential $\mathbf{A}$ is

$$H = \frac{1}{2m}\left|\mathbf{p} - \frac{e}{c}\mathbf{A}\right|^2 + e\phi.$$

The time-dependent Hamilton-Jacobi equation is thus

$$\frac{1}{2m}\left|\nabla S - \frac{e}{c}\mathbf{A}\right|^2 + e\phi = -\frac{\partial S}{\partial t}.$$

(b) Under a gauge transformation of the electromagnetic potentials, the Hamilton-Jacobi equation becomes

$$\frac{1}{2m}\left|\nabla S - \frac{e}{c}\mathbf{A}' + \frac{e}{c}\nabla\lambda\right|^2 + e\phi' + \frac{e}{c}\frac{\partial\lambda}{\partial t} = -\frac{\partial S}{\partial t}.$$

This can be written

$$\frac{1}{2m}\left|\nabla S' - \frac{e}{c}\mathbf{A}'\right|^2 + e\phi' = -\frac{\partial S'}{\partial t}$$

where

$$S' = S + \frac{e}{c}\lambda.$$

The transformed Hamilton-Jacobi equation has the same form as the original equation; the equation is thus invariant under electromagnetic gauge transformations.

Exercise 8.10

(a) Use elementary mechanics to show that the trajectory of a particle of mass m and charge e which moves in a plane (x, y) perpendicular to a uniform magnetic field B is a circle, along which the particle moves with constant angular velocity $\omega = eB/mc$.
(b) Obtain Hamilton's principal function $S_H(x_1, y_1, t_1; x_0, y_0, t_0)$ by integrating the appropriate Lagrangian (in the symmetric gauge)

$$L = \tfrac{1}{2}m(\dot{x}^2 + \dot{y}^2) + \tfrac{1}{2}m\omega(x\dot{y} - y\dot{x})$$

along the path joining the end points.
(Ans. $S_H = \frac{1}{4}m\omega r^2 \cot\frac{1}{2}\omega(t_1 - t_0) + \frac{1}{2}m\omega(x_0 y_1 - x_1 y_0)$ where r is the distance between the end points)

Solution

(a) A particle of mass m and charge e, initially moving perpendicular to a uniform magnetic field **B**, is acted on by a force

$$\mathbf{F} = \frac{e}{c}\mathbf{v} \times \mathbf{B}.$$

This force causes the velocity vector to change. The change is perpendicular to the velocity **v**, so the direction of the velocity changes but its magnitude remains constant.

The change is also perpendicular to **B**, so the particle continues to move perpendicular to **B**, in a plane. The magnitude of the rate of change of **v** is

$$\left|\frac{d\mathbf{v}}{dt}\right| = \frac{F}{m} = \omega v$$

where $\omega = eB/mc$ is the cyclotron frequency. Since B is constant, this rate of change is constant. The particle thus moves in a circle in a plane perpendicular to the magnetic field. The radius of the circle is

$$R = v/\omega .$$

(b) According to part (a), the actual path of a particle, from a start at (x_0, y_0) at time t_0 to a finish at (x_1, y_1) at time t_1, is a circular arc along which the particle moves at constant angular speed ω. The angle swept out, referred to the center of the circle, is

$$\phi = \omega \, \Delta t$$

where $\Delta t = t_1 - t_0$ is the time elapsed between start and finish. The radius R of the circular arc is thus given by (see Fig. 1)

$$2R|\sin \phi/2| = r$$

where $r = \sqrt{(x_1 - x_0)^2 + (y_1 - y_0)^2}$ is the straight line distance between start and finish.

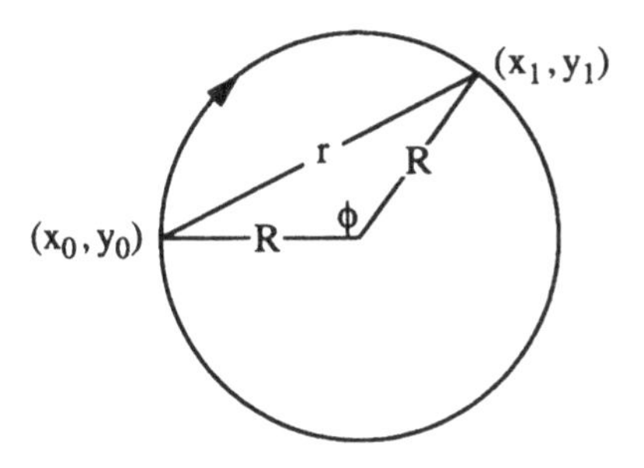

Ex. 8.10, Fig. 1

The linear speed of the particle is

$$v = \omega R = \frac{\omega r}{2|\sin(\omega \, \Delta t/2)|}.$$

Hamilton's principal function is given by

$$S_H = \int_{t_0}^{t} \left[\tfrac{1}{2}mv^2 + \tfrac{1}{2}m\omega(x\dot{y} - y\dot{x})\right]dt$$

where the integration is along the actual path from start to finish. Since the speed v on this path is constant, the first term in S_H becomes

$$\tfrac{1}{2}mv^2\Delta t = \tfrac{1}{2}m\omega R^2\phi.$$

The second term in S_H is

$$\tfrac{1}{2}m\omega\int(x\,dy - y\,dx) = -m\omega \times \text{Area}$$

where "Area" is the area swept out by the radius vector from the origin, positive if the particle moves clockwise and negative if it moves counterclockwise. From Fig. 2 we see that this area is

$$\text{Area} = \tfrac{1}{2}R^2\phi - \tfrac{1}{2}R^2\sin\phi + \tfrac{1}{2}(y_0x_1 - x_0y_1).$$

The first term here is the area of the sector of the circle, the second is (minus) the area of the triangle with vertices at the start, the finish, and the center of the circle, and the third term is the area of the triangle with vertices at the start, the finish, and the origin.

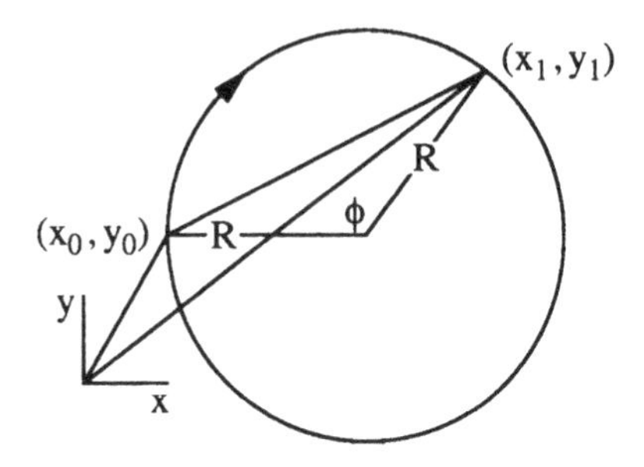

Ex. 8.10, Fig. 2

Putting the pieces together, we obtain Hamilton's principal function

$$\begin{aligned} S_H &= \tfrac{1}{2}m\omega R^2\sin\phi + \tfrac{1}{2}m\omega(x_0y_1 - y_0x_1) \\ &= \tfrac{1}{4}m\omega r^2\cot(\omega\,\Delta t/2) + \tfrac{1}{2}m\omega(x_0y_1 - y_0x_1) \end{aligned}$$

for a charged particle which moves in a plane perpendicular to a uniform magnetic field.

Exercise 8.11

(a) Write down the time-independent Hamilton-Jacobi equation for a particle of mass m and charge e in a uniform magnetic field B in the z-direction. Use cartesian coordinates and a gauge in which the vector potential is $\mathbf{A} = (-By, 0, 0)$.
(b) Show that the variables separate, and obtain a complete integral W.
(c) Use your expression for W to obtain general expressions for the cartesian coordinates as functions of time. Identify physically the separation constants α and their conjugate coordinates β.

Solution

(a) The Hamiltonian for a particle of mass m and charge e in a uniform magnetic field B in the z-direction is

$$H = \frac{1}{2m}\left[\left(p_x + \frac{eB}{c}y\right)^2 + p_y^2 + p_z^2\right].$$

Here we have used cartesian coordinates and a gauge in which the vector potential is $\mathbf{A} = (-By, 0, 0)$. The time-independent Hamilton-Jacobi equation is

$$\frac{1}{2m}\left[\left(\frac{\partial W}{\partial x} + \frac{eB}{c}y\right)^2 + \left(\frac{\partial W}{\partial y}\right)^2 + \left(\frac{\partial W}{\partial z}\right)^2\right] = E.$$

(b) We try a solution of the form

$$W = X(x) + Y(y) + Z(z).$$

The Hamilton-Jacobi equation becomes

$$\frac{1}{2m}\left[\left(\frac{dX}{dx} + \frac{eB}{c}y\right)^2 + \left(\frac{dY}{dy}\right)^2 + \left(\frac{dZ}{dz}\right)^2\right] = E.$$

We rearrange this in the form

$$\left(\frac{dX}{dx} + \frac{eB}{c}y\right)^2 + \left(\frac{dY}{dy}\right)^2 - 2mE = -\left(\frac{dZ}{dz}\right)^2.$$

The left-hand side is a function only of x and y, and the right is a function only of z. Both sides must equal a (negative) constant. We set

$$\left(\frac{dX}{dx} + \frac{eB}{c}y\right)^2 + \left(\frac{dY}{dy}\right)^2 - 2mE = -\alpha_z^2, \qquad \left(\frac{dZ}{dz}\right)^2 = \alpha_z^2.$$

We now rearrange the xy-equation in the form

$$\frac{dX}{dx} = -\frac{eB}{c}y + \sqrt{2mE - \alpha_z^2 - \left(\frac{dY}{dy}\right)^2}.$$

The left-hand side is a function only of x, and the right is a function only of y. Both sides must equal a constant. We set

$$\frac{dX}{dx} = \alpha_x, \qquad \left(\frac{dY}{dy}\right)^2 = 2mE - \alpha_z^2 - \left(\alpha_x + \frac{eB}{c}y\right)^2.$$

The variables are now completely separated, so we can integrate to obtain

$$X = \alpha_x x, \qquad Y = \int\sqrt{2mE - \alpha_z^2 - \left(\alpha_x + (eB/c)y\right)^2}\,dy, \qquad Z = \alpha_z z.$$

The sum of these gives the complete integral

$$W = \alpha_x x + \int\sqrt{2mE - \alpha_z^2 - \left(\alpha_x + (eB/c)y\right)^2}\,dy + \alpha_z z.$$

(c) The complete integral W is the generating function of a canonical transformation. Half of the transformation equations gives the momenta conjugate to x, y, and z,

$$p_x = \frac{\partial W}{\partial x} = \alpha_x, \qquad p_y = \frac{\partial W}{\partial y} = \sqrt{2mE - \alpha_z^2 - \left(\alpha_x + (eB/c)y\right)^2}, \qquad p_z = \frac{\partial W}{\partial z} = \alpha_z.$$

The other half of the transformation equations gives the coordinates conjugate to α_x, α_z, and E,

$$\beta_x = \frac{\partial W}{\partial \alpha_x} = x - \int\frac{\left(\alpha_x + (eB/c)y\right)dy}{\sqrt{2mE - \alpha_z^2 - \left(\alpha_x + (eB/c)y\right)^2}},$$

$$\beta_z = \frac{\partial W}{\partial \alpha_z} = z - \alpha_z\int\frac{dy}{\sqrt{2mE - \alpha_z^2 - \left(\alpha_x + (eB/c)y\right)^2}},$$

$$\beta_E + t = \frac{\partial W}{\partial E} = m\int\frac{dy}{\sqrt{2mE - \alpha_z^2 - \left(\alpha_x + (eB/c)y\right)^2}}.$$

The y-integration in the first of these equations can be performed to give

$$\beta_x = x + \frac{c}{eB}\sqrt{2mE - \alpha_z^2 - \left(\alpha_x + \frac{eB}{c}y\right)^2}\,.$$

This can be rearranged in the form

$$(x - \beta_x)^2 + \left(y + \frac{c\alpha_x}{eB}\right)^2 = \left(\frac{c}{eB}\right)^2\left(2mE - \alpha_z^2\right)$$

and says that the projection of the trajectory onto the xy-plane is a circle with center at $(\beta_x, -c\alpha_x/eB)$ and radius $|c/eB|\sqrt{2mE - \alpha_z^2}$. The second and third equations can be combined to give

$$z(t) = \beta_z + \frac{\alpha_z}{m}(t + \beta_E)$$

and says that the motion in the z-direction is at constant velocity α_z/m. Finally, the y-integration in the third equation can be performed by setting

$$y = -\frac{c\alpha_x}{eB} + \frac{c}{eB}\sqrt{2mE - \alpha_z^2}\sin\Omega, \qquad dy = \frac{c}{eB}\sqrt{2mE - \alpha_z^2}\cos\Omega\, d\Omega.$$

We obtain

$$\Omega(t) = \omega_c(t + \beta_E)$$

where $\omega_c = eB/mc$ is the cyclotron frequency. Substituting back into the yΩ-equation we find

$$y(t) = -\frac{c\alpha_x}{eB} + \frac{c}{eB}\sqrt{2mE - \alpha_z^2}\sin\omega_c(t + \beta_E),$$

and substituting this back into the xy-equation we find

$$x(t) = \beta_x - \frac{c}{eB}\sqrt{2mE - \alpha_z^2}\cos\omega_c(t + \beta_E).$$

These last two equations say that the particle moves around the circle at constant angular speed ω_c with respect to the center of the circle.

Exercise 8.12

(a) Write down the time-independent Hamilton-Jacobi equation for a particle of mass m and charge e in a uniform magnetic field B in the z-direction. Use cylindrical coordinates (ρ, ϕ, z) (the cartesian coordinates are $x = \rho\cos\phi$, $y = \rho\sin\phi$, z) and a gauge in which the vector potential is $\mathbf{A} = \frac{1}{2}B\rho\hat{\boldsymbol{\phi}}$ (its cartesian components are $\mathbf{A} = (-\frac{1}{2}By, +\frac{1}{2}Bx, 0)$).
(b) Show that the variables separate, and obtain a complete integral W.
(c) Use your expression for W to obtain general expressions for the cylindrical coordinates as functions of time. Identify physically the separation constants α and their conjugate coordinates β.

Solution

(a) The time-independent Hamilton-Jacobi equation for a particle of mass m and charge e in a uniform magnetic field in the z-direction is

$$\frac{1}{2m}\left|\nabla W - \frac{e}{c}\mathbf{A}\right|^2 = E.$$

Here **A** is the vector potential. If we use cylindrical coordinates (ρ, ϕ, z) so that

$$\nabla W = \frac{\partial W}{\partial \rho}\hat{\boldsymbol{\rho}} + \frac{1}{\rho}\frac{\partial W}{\partial \phi}\hat{\boldsymbol{\phi}} + \frac{\partial W}{\partial z}\hat{\mathbf{z}}$$

and choose a gauge in which the vector potential is

$$\mathbf{A} = \tfrac{1}{2}B\rho\hat{\boldsymbol{\phi}},$$

the Hamilton-Jacobi equation becomes

$$\frac{1}{2m}\left[\left(\frac{\partial W}{\partial \rho}\right)^2 + \left(\frac{1}{\rho}\frac{\partial W}{\partial \phi} - \frac{eB}{2c}\rho\right)^2 + \left(\frac{\partial W}{\partial z}\right)^2\right] = E.$$

(b) We try a solution of the form

$$W = R(\rho) + \Phi(\phi) + Z(z).$$

The Hamilton-Jacobi equation can be rearranged in the form

$$\left(\frac{dR}{d\rho}\right)^2 + \left(\frac{1}{\rho}\frac{d\Phi}{d\phi} - \frac{eB}{2c}\rho\right)^2 - 2mE = -\left(\frac{dZ}{dz}\right)^2.$$

The left-hand side is a function only of ρ and ϕ, and the right-hand side is a function only of z. Both sides must equal a constant. We set

$$\left(\frac{dR}{d\rho}\right)^2 + \left(\frac{1}{\rho}\frac{d\Phi}{d\phi} - \frac{eB}{2c}\rho\right)^2 - 2mE = -\alpha_z^2, \qquad \left(\frac{dZ}{dz}\right)^2 = \alpha_z^2.$$

The $\rho\phi$-equation can be rearranged in the form

$$\frac{d\Phi}{d\phi} = \frac{eB}{2c}\rho^2 + \rho\sqrt{2mE - \alpha_z^2 - \left(\frac{dR}{d\rho}\right)^2}.$$

The left-hand side is a function only of ϕ, and the right-hand side is a function only of ρ. Both sides must equal a constant. We set

$$\frac{d\Phi}{d\phi} = \alpha_\phi, \qquad \frac{eB}{2c}\rho^2 + \rho\sqrt{2mE - \alpha_z^2 - \left(\frac{dR}{d\rho}\right)^2} = \alpha_\phi.$$

The ρ-equation can be written

$$\left(\frac{dR}{d\rho}\right)^2 = 2mE - \alpha_z^2 - \left(\frac{\alpha_\phi}{\rho} - \frac{eB}{2c}\rho\right)^2.$$

The variables are now completely separated, and we can integrate to obtain R, Φ, and Z. Their sum gives the complete integral

$$W = \int\sqrt{2mE - \alpha_z^2 - \left(\frac{\alpha_\phi}{\rho} - \frac{eB}{2c}\rho\right)^2}\, d\rho + \alpha_\phi\phi + \alpha_z z.$$

(c) The complete integral W is the generating function of a canonical transformation. The first half of the transformation equations gives the momenta conjugate to ρ, ϕ, and z,

$$p_\rho = \frac{\partial W}{\partial \rho} = \sqrt{2mE - \alpha_z^2 - \left(\frac{\alpha_\phi}{\rho} - \frac{eB}{2c}\rho\right)^2}, \quad p_\phi = \frac{\partial W}{\partial \phi} = \alpha_\phi, \quad p_z = \frac{\partial W}{\partial z} = \alpha_z.$$

We see that the separation constants α_ϕ and α_z are the constant values of the momenta conjugate to ϕ and to z. The second half of the transformation equations gives the coordinates conjugate to α_ϕ, α_z, and E,

$$\beta_\phi = \frac{\partial W}{\partial \alpha_\phi} = \int \frac{-\left((\alpha_\phi/\rho^2) - (eB/2c)\right)d\rho}{\sqrt{2mE - \alpha_z^2 - \left((\alpha_\phi/\rho) - (eB/2c)\rho\right)^2}} + \phi,$$

$$\beta_z = \frac{\partial W}{\partial \alpha_z} = -\alpha_z \int \frac{d\rho}{\sqrt{2mE - \alpha_z^2 - \left((\alpha_\phi/\rho) - (eB/2c)\rho\right)^2}} + z,$$

$$\beta_E + t = \frac{\partial W}{\partial E} = m \int \frac{d\rho}{\sqrt{2mE - \alpha_z^2 - \left((\alpha_\phi/\rho) - (eB/2c)\rho\right)^2}}.$$

To do the ρ-integration in the first of these equations, we let

$$u = \frac{\alpha_\phi}{\rho} + \frac{eB}{2c}\rho, \quad u^2 = \left(\frac{\alpha_\phi}{\rho} - \frac{eB}{2c}\rho\right)^2 + 2\frac{eB}{c}\alpha_\phi, \quad du = -\left(\frac{\alpha_\phi}{\rho^2} - \frac{eB}{2c}\right)d\rho,$$

so the equation becomes

$$\phi - \beta_\phi = \int \frac{-du}{\sqrt{2mE - \alpha_z^2 + 2(eB/c)\alpha_\phi - u^2}}.$$

The integration is now simple, and we find the orbit equation

$$(u =)\ \frac{\alpha_\phi}{\rho} + \frac{eB}{2c}\rho = \sqrt{2mE - \alpha_z^2 + 2\frac{eB}{c}\alpha_\phi}\cos(\phi - \beta_\phi).$$

This is the equation of a circle. To see this, we apply the cosine law to the triangle in Fig. 1 and find

$$R^2 = \rho^2 + \rho_0^2 - 2\rho\rho_0\cos(\phi - \phi_0)$$

where (ρ_0, ϕ_0) are the polar coordinates of the center of the circle and R is its radius. This can be rearranged in the form

$$\frac{\rho_0^2 - R^2}{\rho} + \rho = 2\rho_0\cos(\phi - \phi_0)$$

and has the same form as the orbit equation, with

$$\alpha_\phi = \frac{eB}{2c}\left(\rho_0^2 - R^2\right), \quad \sqrt{2mE - \alpha_z^2 + 2\frac{eB}{c}\alpha_\phi} = \frac{eB}{c}\rho_0, \quad \text{and} \quad \beta_\phi = \phi_0.$$

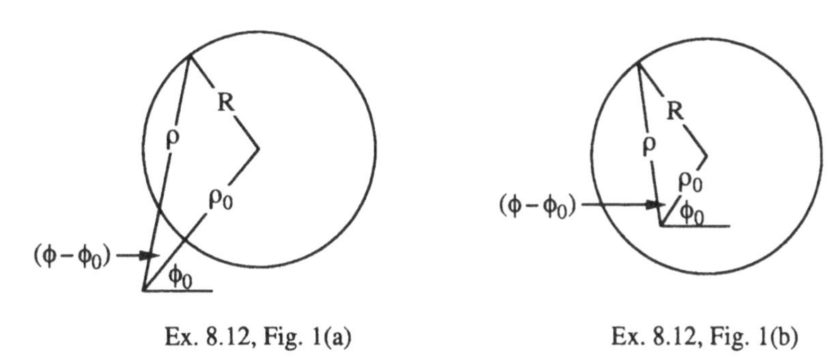

Ex. 8.12, Fig. 1(a) Ex. 8.12, Fig. 1(b)

If α_ϕ is positive, $\rho_0 > R$ and the circle does not enclose the origin (Fig. 1(a)). If α_ϕ is negative, $\rho_0 < R$ and the circle encloses the origin (Fig. 1(b)). The minimum and maximum distances to the origin are $\rho_{min} = |\rho_0 - R|$ and $\rho_{max} = \rho_0 + R$. The radius of the circle is $R = \frac{c}{eB}\sqrt{2mE - \alpha_z^2}$; the square root here is the component of the linear momentum perpendicular to the magnetic field. The coordinate β_ϕ gives the direction of the center of the circle.

To find out how the particle moves around the circle in time, we consider the third transformation equation, which we write

$$\beta_E + t = m\int \frac{\rho\, d\rho}{\sqrt{(2mE - \alpha_z^2)\rho^2 - \left(\alpha_\phi - (eB/2c)\rho^2\right)^2}}$$

$$= \frac{mc}{eB}\int \frac{2\rho\, d\rho}{\sqrt{4R^2\rho^2 - \left(\rho_0^2 - R^2 - \rho^2\right)^2}} = \frac{mc}{eB}\int \frac{2\rho\, d\rho}{\sqrt{4R^2\rho_0^2 - \left(\rho_0^2 + R^2 - \rho^2\right)^2}}.$$

We now set

$$\rho^2 = \rho_0^2 + R^2 - 2\rho_0 R\cos\Omega.$$

The variable Ω is the angular position of the particle referred to the center of the circle, with $\Omega = 0$ at $\rho = \rho_{min}$ (see Fig. 2). Integration gives

$$\Omega = \omega_c(t + \beta_E)$$

where $\omega_c = eB/mc$ is the cyclotron frequency. The particle thus moves around the circle at constant angular speed ω_c.

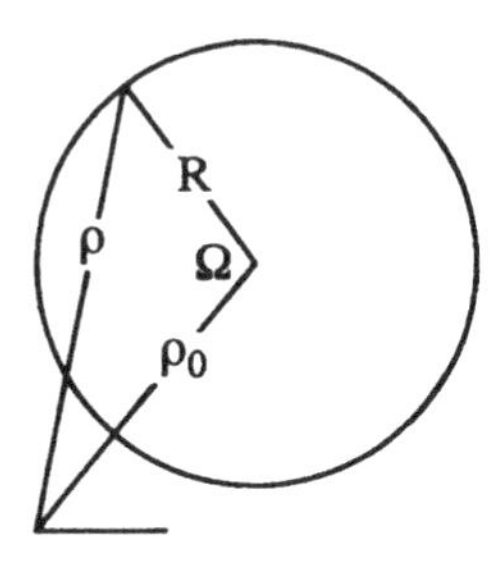

Ex. 8.12, Fig. 2

Finally, we note that the integrals in the second and third transformation equations are the same, so these equations can be combined to give the motion in the z-direction

$$z(t) = \beta_z + \frac{\alpha_z}{m}(t + \beta_E).$$

The motion in the z-direction is at constant velocity α_z/m. The coordinate β_z is the "initial" $(t + \beta_E = 0)$ value of z.

Exercise 8.13

A particle of mass m and charge e moves in uniform crossed electric and magnetic fields, $\mathcal{E}$ in the x-direction and $\mathcal{B}$ in the z-direction.
(a) Write down the time-independent Hamilton-Jacobi equation in cartesian coordinates, and show that the variables separate for a suitable choice of gauge for the electromagnetic potentials.
(b) Use your solution to find general expressions for the cartesian coordinates of the particle as functions of time.

Solution

The Hamilton-Jacobi equation for a particle of mass m and charge e in an electromagnetic field which consists of a uniform electric field $\mathcal{E}$ in the x-direction and a uniform magnetic field $\mathcal{B}$ in the z-direction is

$$\frac{1}{2m}\left(\left(\frac{\partial W}{\partial x}\right)^2 + \left(\frac{\partial W}{\partial y} - \frac{e\mathcal{B}}{c}x\right)^2 + \left(\frac{\partial W}{\partial z}\right)^2\right) - e\mathcal{E}x = E.$$

Here we have chosen a gauge in which the vector potential is $\mathbf{A} = (0, \mathcal{B}x, 0)$ and the scalar potential is $\phi = -\mathcal{E}x$. The advantage of this choice is that it allows separation of variables, as we now show. We try a solution of the form

$$W = X(x) + Y(y) + Z(z).$$

The Hamilton-Jacobi equation becomes

$$\left(\frac{dX}{dx}\right)^2 + \left(\frac{dY}{dy} - \frac{e\mathcal{B}}{c}x\right)^2 - 2me\mathcal{E}x - 2mE = -\left(\frac{dZ}{dz}\right)^2.$$

The left-hand side is a function only of x and y while the right-hand side is a function only of z. Both sides must equal a constant. We set

$$\left(\frac{dX}{dx}\right)^2 + \left(\frac{dY}{dy} - \frac{e\mathcal{B}}{c}x\right)^2 - 2me\mathcal{E}x - 2mE = -\alpha_z^2, \qquad \left(\frac{dZ}{dz}\right)^2 = \alpha_z^2.$$

We now rearrange the xy-equation in the form

$$\frac{dY}{dy} = \frac{e\mathcal{B}}{c}x + \sqrt{2mE - \alpha_z^2 + 2me\mathcal{E}x - \left(\frac{dX}{dx}\right)^2}.$$

The left-hand side is a function only of y and the right-hand side is a function only of x. Both sides must equal a constant. We set

$$\frac{e\mathcal{B}}{c}x + \sqrt{2mE - \alpha_z^2 + 2me\mathcal{E}x - \left(\frac{dX}{dx}\right)^2} = \alpha_y, \qquad \frac{dY}{dy} = \alpha_y.$$

The x-equation can be written

$$\left(\frac{dX}{dx}\right)^2 = 2mE - \alpha_z^2 + 2me\mathcal{E}x - \left(\alpha_y - \frac{e\mathcal{B}}{c}x\right)^2.$$

The variables are now completely separated, and we can integrate to obtain X, Y, and Z. Their sum gives the complete integral

$$W = \int \sqrt{2mE - \alpha_z^2 + 2me\mathcal{E}x - \left(\alpha_y - (e\mathcal{B}/c)x\right)^2}\, dx + \alpha_y y + \alpha_z z.$$

The canonical transformation generated by W gives

$$\beta_y = \frac{\partial W}{\partial \alpha_y} = \int \frac{-\left(\alpha_y - (e\mathcal{B}/c)x\right)dx}{\sqrt{2mE - \alpha_z^2 + 2me\mathcal{E}x - \left(\alpha_y - (e\mathcal{B}/c)x\right)^2}} + y,$$

$$\beta_z = \frac{\partial W}{\partial \alpha_z} = -\alpha_z \int \frac{dx}{\sqrt{2mE - \alpha_z^2 + 2me\mathcal{E}x - \left(\alpha_y - (e\mathcal{B}/c)x\right)^2}} + z,$$

$$\beta_E + t = \frac{\partial W}{\partial E} = m \int \frac{dx}{\sqrt{2mE - \alpha_z^2 + 2me\mathcal{E}x - \left(\alpha_y - (e\mathcal{B}/c)x\right)^2}}.$$

The first of these equations can be written

$$y - \beta_y = -\int \frac{\left(x - (1/m\omega)\alpha_y\right)dx}{\sqrt{a^2 - \left(x - (1/m\omega)(\alpha_y + e\mathcal{E}/\omega)\right)^2}}$$

where $\omega = e\mathcal{B}/mc$ is the cyclotron frequency and

$$(m\omega a)^2 = 2mE - \alpha_z^2 + 2(e\mathcal{E}/\omega)\alpha_y + (e\mathcal{E}/\omega)^2.$$

To perform the x-integration, we set

$$x = \left(\frac{\alpha_y}{m\omega} + \frac{e\mathcal{E}}{m\omega^2}\right) - a\cos\Omega, \qquad dx = a\sin\Omega\, d\Omega.$$

The xy-equation becomes

$$y - \beta_y = -\int \left(\frac{e\mathcal{E}}{m\omega^2} - a\cos\Omega\right) d\Omega = -\frac{e\mathcal{E}}{m\omega^2}\Omega + a\sin\Omega.$$

These last two equations, which express x and y in terms of the parameter Ω, give the projection of the trajectory onto the xy-plane. We recognize the curve as a cycloid. To find how the particle moves along the trajectory, we turn to the third transformation equation, which we write

$$\omega(t + \beta_E) = \int \frac{dx}{\sqrt{a^2 - \left(x - (1/m\omega)(\alpha_y + e\mathcal{E}/\omega)\right)^2}} = \int d\Omega = \Omega(t).$$

The parameter Ω thus increases uniformly at a rate ω, and the x and y coordinates, as functions of time, are

$$x(t) = \left(\frac{\alpha_y}{m\omega} + \frac{e\mathcal{E}}{m\omega^2}\right) - a\cos\omega(t + \beta_E),$$

$$y(t) = \beta_y - \frac{e\mathcal{E}}{m\omega}(t + \beta_E) + a\sin\omega(t + \beta_E).$$

Finally, the z-coordinate, as a function of time, can be found by combining the second and third transformation equations, to give

$$z(t) = \beta_z + \frac{\alpha_z}{m}(t + \beta_E).$$

Motion in the z-direction is at constant velocity α_z/m.

CHAPTER IX

ACTION-ANGLE VARIABLES

Exercise 9.01

A particle of mass m moves in one dimension x in a potential well

$$V = V_0 \tan^2(\pi x/2a)$$

where V_0 and a are constants. Find the action variable I, express the total energy E in terms of I, and find the frequency $\omega = dE/dI$. In particular examine and interpret the low energy ($E \ll V_0$) and high energy ($E \gg V_0$) limits of your expressions (refer to Exercise 1.01).

Solution

The momentum of the particle, at position x, is

$$p = \sqrt{2m\left(E - V_0 \tan^2 \frac{\pi x}{2a}\right)}.$$

The particle oscillates back and forth between limits $\pm A$ where

$$\tan \frac{\pi A}{2a} = \sqrt{\frac{E}{V_0}}.$$

The action variable I is given by

$$I = \frac{4}{2\pi} \int_0^A \sqrt{2m\left(E - V_0 \tan^2 \frac{\pi x}{2a}\right)}\, dx$$

$$= \frac{4a}{\pi^2} \sqrt{2m} \left[\sqrt{E + V_0} \int_0^{\theta_{max}} \frac{\cos\theta \, d\theta}{\sqrt{\dfrac{E}{E + V_0} - \sin^2\theta}} - \sqrt{V_0} \int_0^{\theta_{max}} \frac{d\theta/\cos^2\theta}{\sqrt{\dfrac{E}{V_0} - \tan^2\theta}} \right]$$

where $\theta = \pi x/2a$ and $\theta_{max} = \pi A/2a$. The first integral can be performed with the substitution

$$\sin\theta = \sqrt{\frac{E}{E + V_0}} \sin\alpha, \qquad \cos\theta\, d\theta = \sqrt{\frac{E}{E + V_0}} \cos\alpha\, d\alpha,$$

and the second can be performed with the substitution

$$\tan\theta = \sqrt{\frac{E}{V_0}}\sin\beta, \qquad \frac{d\theta}{\cos^2\theta} = \sqrt{\frac{E}{V_0}}\cos\beta\, d\beta.$$

The result is

$$I = \frac{2a}{\pi}\sqrt{2m}\left[\sqrt{E+V_0} - \sqrt{V_0}\right].$$

Solving for the energy, we find

$$E = \left(\sqrt{V_0} + \frac{\pi I}{2a\sqrt{2m}}\right)^2 - V_0 = \frac{\pi}{a}\sqrt{\frac{V_0}{2m}}\, I + \frac{\pi^2}{8ma^2} I^2.$$

The angular frequency of oscillation is

$$\omega = \frac{\partial E}{\partial I} = \frac{\pi}{a}\sqrt{\frac{V_0}{2m}} + \frac{\pi^2}{4ma^2} I.$$

At low energy ($E << V_0$) the action variable and angular frequency become

$$I = \frac{2a}{\pi}\sqrt{2mV_0}\left[\sqrt{1+\frac{E}{V_0}} - 1\right] \approx \frac{a}{\pi}\sqrt{\frac{2m}{V_0}}E + \cdots \quad \text{and} \quad \omega \approx \frac{\pi}{a}\sqrt{\frac{V_0}{2m}}.$$

Also, the potential becomes

$$V(x) \approx V_0\left(\frac{\pi x}{2a}\right)^2 = \frac{1}{2}\left(\frac{\pi^2 V_0}{2a^2}\right)x^2.$$

In this limit the system behaves like a simple harmonic oscillator with spring constant $\pi^2 V_0/2a^2$. At high energy ($E >> V_0$) the action variable and angular frequency become

$$I \approx \frac{4a}{2\pi}\sqrt{2mE} \quad \text{and} \quad \omega \approx \frac{2\pi}{4a}\sqrt{\frac{2E}{m}}.$$

In this limit the system behaves like a particle in an infinite square well potential ($V = 0$ for $|x| < a$ and $V \to \infty$ for $|x| > a$).

Exercise 9.02

A particle of mass m moves in two dimensions (x, y) in a non-isotropic simple harmonic oscillator well

$$V(x,y) = \tfrac{1}{2}m\omega_x^2 x^2 + \tfrac{1}{2}m\omega_y^2 y^2$$

where in general $\omega_x \neq \omega_y$.

(a) Find the action variables (I_x, I_y), and express the energy in terms of these.

(b) Find the angle variables (ϕ_x, ϕ_y), and express the cartesian coordinates in terms of the action-angle variables.

(c) Write down the angle variables and the cartesian coordinates as functions of time.

(d) Sketch the trajectories of the particle in (x, y) space and in (ϕ_x, ϕ_y) space.

Solution

(a) The Hamiltonian is

$$H = \frac{1}{2m}\left[p_x^2 + p_y^2\right] + \frac{1}{2}m\omega_x^2 x^2 + \frac{1}{2}m\omega_y^2 y^2,$$

so the time-independent Hamilton-Jacobi equation is

$$\frac{1}{2m}\left[\left(\frac{\partial W}{\partial x}\right)^2 + \left(\frac{\partial W}{\partial y}\right)^2\right] + \frac{1}{2}m\omega_x^2 x^2 + \frac{1}{2}m\omega_y^2 y^2 = E.$$

We try a solution of the form

$$W = X(x) + Y(y).$$

The Hamilton-Jacobi equation becomes

$$\left[\frac{1}{2m}\left(\frac{dX}{dx}\right)^2 + \frac{1}{2}m\omega_x^2 x^2\right] + \left[\frac{1}{2m}\left(\frac{dY}{dy}\right)^2 + \frac{1}{2}m\omega_y^2 y^2\right] = E.$$

The first term on the left is a function only of x, the second term is a function only of y, and their sum is the constant E. Both terms must equal a constant. We set

$$\frac{1}{2m}\left(\frac{dX}{dx}\right)^2 + \frac{1}{2}m\omega_x^2 x^2 = \alpha_x, \qquad \frac{1}{2m}\left(\frac{dY}{dy}\right)^2 + \frac{1}{2}m\omega_y^2 y^2 = \alpha_y,$$

where α_x and α_y are separation constants with $\alpha_x + \alpha_y = E$. Integration gives

$$X = \int \sqrt{2m\alpha_x - m^2\omega_x^2 x^2}\, dx, \qquad Y = \int \sqrt{2m\alpha_y - m^2\omega_y^2 y^2}\, dy,$$

so a complete integral to the Hamilton-Jacobi equation is

$$W = \int \sqrt{2m\alpha_x - m^2\omega_x^2 x^2}\, dx + \int \sqrt{2m\alpha_y - m^2\omega_y^2 y^2}\, dy.$$

The first half of the canonical transformation generated by W gives the momenta conjugate to x and y,

$$p_x = \frac{\partial W}{\partial x} = \sqrt{2m\alpha_x - m^2\omega_x^2 x^2}, \qquad p_y = \frac{\partial W}{\partial y} = \sqrt{2m\alpha_y - m^2\omega_y^2 y^2}.$$

The action variables are then

$$I_x = \frac{1}{2\pi}\oint \sqrt{2m\alpha_x - m^2\omega_x^2 x^2}\, dx, \qquad I_y = \frac{1}{2\pi}\oint \sqrt{2m\alpha_y - m^2\omega_y^2 y^2}\, dy.$$

Each of these integrals is the same as that for a one-dimensional oscillator, so we find

$$I_x = \alpha_x/\omega_x, \qquad I_y = \alpha_y/\omega_y.$$

We can turn these around to express the separation constants in terms of the action variables,

$$\alpha_x = \omega_x I_x, \qquad \alpha_y = \omega_y I_y.$$

The total energy is then

$$E = \omega_x I_x + \omega_y I_y.$$

The Jacobi complete integral becomes

$$W = \int \sqrt{2m\omega_x I_x - m^2\omega_x^2 x^2}\, dx + \int \sqrt{2m\omega_y I_y - m^2\omega_y^2 y^2}\, dy.$$

(b) The second half of the canonical transformation generated by W gives the angle variables ϕ_x and ϕ_y conjugate to the action variables I_x and I_y,

$$\phi_x = \frac{\partial W}{\partial I_x} = \int \frac{dx}{\sqrt{(2I_x/m\omega_x) - x^2}}, \qquad \phi_y = \frac{\partial W}{\partial I_y} = \int \frac{dy}{\sqrt{(2I_y/m\omega_y) - y^2}}.$$

Integration proceeds as for a one-dimensional oscillator, and we find that the angle variables ϕ_x and ϕ_y are the phase angles with

$$x = \sqrt{\frac{2I_x}{m\omega_x}}\sin\phi_x \quad \text{and} \quad y = \sqrt{\frac{2I_y}{m\omega_y}}\sin\phi_y .$$

(c) Hamilton's equations for the action-angle variables are

$$\frac{d\phi_x}{dt} = \frac{\partial E}{\partial I_x} = \omega_x, \qquad \frac{d\phi_y}{dt} = \frac{\partial E}{\partial I_y} = \omega_y,$$

$$\frac{dI_x}{dt} = -\frac{\partial E}{\partial \phi_x} = 0, \qquad \frac{dI_y}{dt} = -\frac{\partial E}{\partial \phi_y} = 0.$$

The action variables are constant in time, and the angle variables increase uniformly with time,

$$\phi_x(t) = \phi_{x0} + \omega_x t, \qquad \phi_y(t) = \phi_{y0} + \omega_y t.$$

The cartesian coordinates, as functions of time, are then given by the canonical transformation

$$x(t) = \sqrt{\frac{2I_x}{m\omega_x}}\sin(\omega_x t + \phi_{x0}), \qquad y(t) = \sqrt{\frac{2I_y}{m\omega_y}}\sin(\omega_y t + \phi_{y0}).$$

(d) The resulting trajectories in (x,y) space are Lissajous figures within the rectangle $|x| < \sqrt{2I_x/m\omega_x}$, $|y| < \sqrt{2I_y/m\omega_y}$. See Fig. 1(a), in which the frequency ratio has been taken to be $\omega_x/\omega_y = 5/3$. Since this is a rational number, this trajectory is periodic, closing after five cycles of x and three of y.

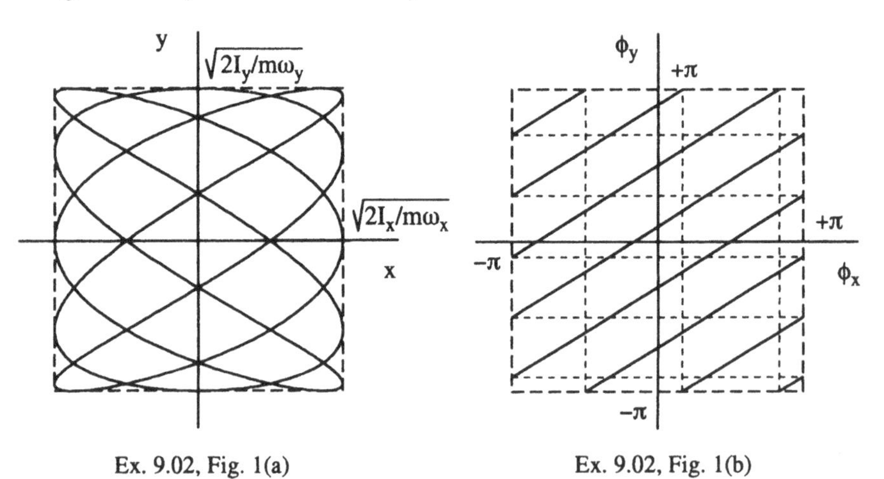

Ex. 9.02, Fig. 1(a)

Ex. 9.02, Fig. 1(b)

The trajectories in (ϕ_x, ϕ_y) cartesian space are straight lines. Since, however, the phase angles $\phi - \pi$ and $\phi + \pi$ are equivalent, we can confine our attention to the square $|\phi_x| < \pi$, $|\phi_y| < \pi$. Whenever a trajectory leaves the square, it is continued inside by translation through 2π. See Fig. 1(b), which displays the same trajectory for the same physical system as Fig. 1(a). We can imagine the boundaries $\phi_x = \pm\pi$ joined and then the boundaries $\phi_y = \pm\pi$ joined. The result is a torus around which the trajectory wraps.

Exercise 9.03

A particle of mass m moves in two dimensions (x, y) in a rectangular "infinite square well" potential (sometimes called a rectangular billiard)

$$V = 0 \text{ for } 0 < x < a,\ 0 < y < b \quad \text{and} \quad V \to \infty \text{ otherwise.}$$

(a) Find the action variables I_x and I_y.
(b) Find the frequencies ω_x and ω_y, and write down the condition for periodic trajectories. Interpret your result geometrically.

Solution

(a) The Hamiltonian is

$$H = \frac{1}{2m}\left[p_x^2 + p_y^2\right] \quad \text{with} \quad 0 < x < a,\ 0 < y < b.$$

The Hamilton-Jacobi equation is

$$\frac{1}{2m}\left[\left(\frac{\partial W}{\partial x}\right)^2 + \left(\frac{\partial W}{\partial y}\right)^2\right] = E.$$

We try a solution of the form

$$W = X(x) + Y(y).$$

The Hamilton-Jacobi equation becomes

$$\left(\frac{dX}{dx}\right)^2 + \left(\frac{dY}{dy}\right)^2 = 2mE.$$

The first term on the left is a function only of x, the second term is a function only of y, and their sum is the constant $2mE$. Both terms must equal a constant. We set

$$\left(\frac{dX}{dx}\right)^2 = \alpha_x^2, \qquad \left(\frac{dY}{dy}\right)^2 = \alpha_y^2,$$

where α_x and α_y are separation constants (which we take to be positive) with

$$E = \frac{1}{2m}\left[\alpha_x^2 + \alpha_y^2\right].$$

Integration gives $X = \pm\alpha_x x$ and $Y = \pm\alpha_y y$, and their sum gives the complete integral

$$W = \pm\alpha_x x \pm \alpha_y y,$$

a function of four branches. The first half of the canonical transformation generated by W gives the momenta conjugate to x and y,

$$p_x = \frac{\partial W}{\partial x} = \pm\alpha_x, \qquad p_y = \frac{\partial W}{\partial y} = \pm\alpha_y.$$

Whenever the trajectory hits a boundary, the corresponding momentum changes sign. The action variables are

$$I_x = \frac{1}{2\pi}\oint p_x\,dx = \frac{1}{2\pi}\int_0^a \alpha_x\,dx + \frac{1}{2\pi}\int_a^0 (-\alpha_x)\,dx = \frac{\alpha_x a}{\pi},$$
$$I_y = \frac{1}{2\pi}\oint p_y\,dy = \frac{1}{2\pi}\int_0^b \alpha_y\,dy + \frac{1}{2\pi}\int_b^0 (-\alpha_y)\,dy = \frac{\alpha_y b}{\pi}.$$

We can turn these around to express the separation constants in terms of the action variables,

$$\alpha_x = \frac{\pi I_x}{a}, \qquad \alpha_y = \frac{\pi I_y}{b}.$$

The energy is then

$$E = \frac{\pi^2}{2m}\left[\frac{I_x^2}{a^2} + \frac{I_y^2}{b^2}\right].$$

(b) The frequencies associated with the two degrees of freedom are

$$\omega_x = \frac{\partial E}{\partial I_x} = \frac{\pi^2 I_x}{ma^2}, \qquad \omega_y = \frac{\partial E}{\partial I_y} = \frac{\pi^2 I_y}{mb^2}.$$

These are easily understood: the magnitude of the velocity of the particle in the x-direction is $|v_x| = |p_x|/m = \pi I_x/ma$, and the time it takes for the particle to return to its initial x-position with its initial x-velocity is $\Delta t_x = 2a/|v_x| = 2ma^2/\pi I_x$; the angular frequency for the x-motion is thus $\omega_x = 2\pi/\Delta t_x = \pi^2 I_x/ma^2$; similarly for the y-motion. Periodic orbits occur for those trajectories for which $\omega_x/\omega_y = I_x b^2/I_y a^2 = n_x/n_y$ where n_x and n_y are integers. They are the trajectories for which the particle returns to its initial position with its initial velocity after a time $\Delta t = n_x \Delta t_x = n_y \Delta t_y$ (that is, after n_x cycles of x and n_y of y).

Exercise 9.04

A particle of mass m moves in two dimensions (ρ, ϕ) in a circular "infinite square well" potential (sometimes called a circular billiard)

$$V = 0 \text{ for } \rho < a \quad \text{and} \quad V \to \infty \text{ for } \rho \geq a.$$

(a) Find the action variables I_ρ and I_ϕ.
(b) Find the frequencies ω_ρ and ω_ϕ, and write down the condition for periodic trajectories. Interpret your result geometrically.

Solution

According to Exercise 8.05, the momenta conjugate to the plane polar coordinates ρ and ϕ are given by

$$p_\rho = \sqrt{2mE - L^2/\rho^2}, \qquad p_\phi = L,$$

where E is the (kinetic) energy and L is the angular momentum. The radial coordinate is oscillatory, with ρ oscillating between an inner turning radius

$$b = \frac{|L|}{\sqrt{2mE}}$$

and an outer radius a. The angular coordinate ϕ is rotational, with ϕ increasing by 2π per cycle for L positive and decreasing by 2π for L negative.

The action variable for the radial degree of freedom is

$$I_\rho = \frac{1}{\pi}\int_b^a \sqrt{2mE - L^2/\rho^2}\, d\rho$$
$$= \frac{|L|}{\pi b}\int_b^a \frac{\rho\, d\rho}{\sqrt{\rho^2 - b^2}} - \frac{|L|}{\pi}\int_b^a \frac{d\rho/\rho^2}{\sqrt{1/b^2 - 1/\rho^2}}$$
$$= \frac{|L|}{\pi}(\tan\alpha - \alpha)$$

where $\cos\alpha = b/a$. The parameter α is the angle shown in Fig. 1.

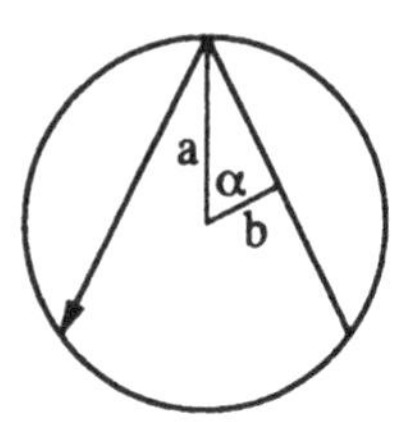

Ex. 9.04, Fig. 1

The action variable for the angular degree of freedom is

$$I_\phi = |L|.$$

To express the energy E in terms of the action variables, we set

$$E = \frac{L^2}{2mb^2} = \frac{I_\phi^2}{2ma^2\cos^2\alpha}$$

and determine the angle α from

$$\tan\alpha - \alpha = \pi I_\rho / I_\phi .$$

The frequency for the radial degree of freedom is

$$\omega_\rho = \frac{\partial E}{\partial I_\rho} = \frac{I_\phi^2 \sin\alpha}{ma^2\cos^3\alpha}\left(\frac{\partial\alpha}{\partial I_\rho}\right) = \frac{2\pi I_\phi}{ma^2\sin 2\alpha},$$

where we have used $\tan^2\alpha\left(\frac{\partial\alpha}{\partial I_\rho}\right) = \frac{\pi}{I_\phi}$. Since $I_\phi = |L| = b\sqrt{2mE} = a\cos\alpha\sqrt{2mE}$, the radial frequency can be written

$$\omega_\rho = 2\pi\frac{\sqrt{2E/m}}{2a\sin\alpha}.$$

To understand this, note that $\sqrt{2E/m}$ is the constant speed of the particle and $2a\sin\alpha$ is the distance gone in one cycle of ρ, so $\Delta t_\rho = \dfrac{2a\sin\alpha}{\sqrt{2E/m}}$ is the time for one cycle of ρ and $\omega_\rho = 2\pi/\Delta t_\rho$.

The frequency for the angular degree of freedom is

$$\begin{aligned}\omega_\phi &= \frac{\partial E}{\partial I_\phi} = \frac{I_\phi}{ma^2\cos^2\alpha} + \frac{I_\phi^2\sin\alpha}{ma^2\cos^3\alpha}\left(\frac{\partial\alpha}{\partial I_\phi}\right)\\ &= \frac{I_\phi}{ma^2\cos^2\alpha} - \frac{\pi I_\rho}{ma^2\sin\alpha\cos\alpha}\\ &= \frac{2\alpha I_\phi}{ma^2\sin 2\alpha}\end{aligned}$$

where we have used $\tan^2\alpha\left(\dfrac{\partial\alpha}{\partial I_\phi}\right) = -\dfrac{\pi I_\rho}{I_\phi^2}$.

Closed periodic orbits occur for those values of the action variables for which

$$\frac{\omega_\phi}{\omega_\rho} = \frac{2\alpha}{2\pi} = \frac{n_\phi}{n_\rho}$$

where n_ϕ and n_ρ are integers, with $n_\phi/n_\rho \le 1/2$ since $\alpha \le \pi/2$. In one cycle of the radial coordinate the angular coordinate advances by 2α, and in n_ρ cycles it advances by $2\alpha n_\rho$. If this equals an integral multiple of 2π, $2\pi n_\phi$ say, the orbit returns to its start and is thus a closed periodic orbit. Examples of periodic orbits are shown in Fig. 2.

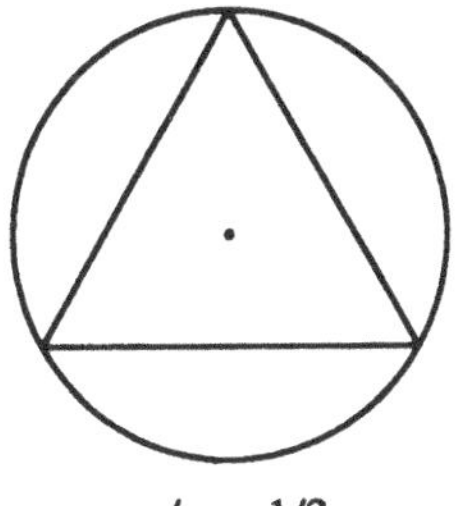

$n_\phi/n_\rho = 1/3$

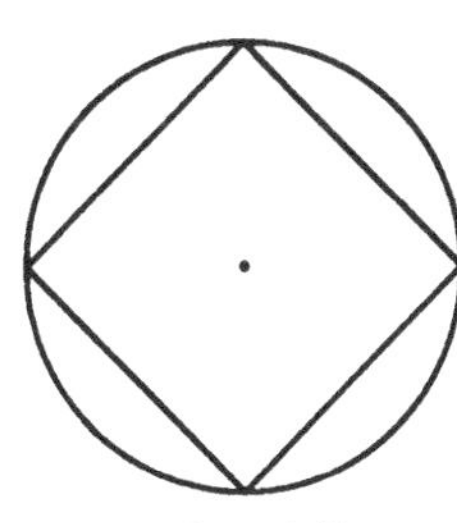

$n_\phi/n_\rho = 1/4$

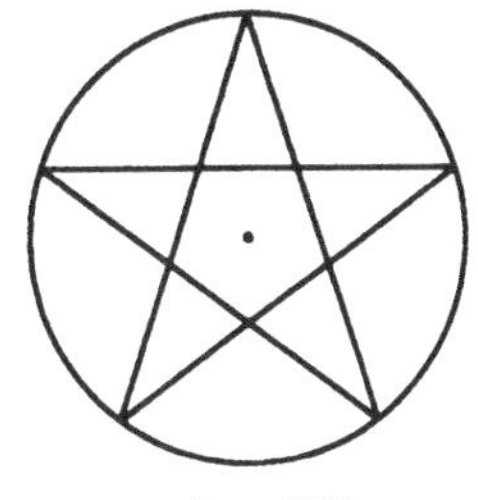

$n_\phi/n_\rho = 2/5$

Ex. 9.04, Fig. 2

Exercise 9.05

A particle of mass m moves in a three-dimensional isotropic oscillator well

$$V = \tfrac{1}{2}m\omega^2(x^2 + y^2 + z^2) = \tfrac{1}{2}m\omega^2(\rho^2 + z^2) = \tfrac{1}{2}m\omega^2 r^2.$$

(a) Separate the Hamilton-Jacobi equation in cartesian coordinates (x, y, z), find the action variables, and express the Hamiltonian in terms of these. Find the frequencies $(\omega_x, \omega_y, \omega_z)$.
(b) Separate the Hamilton-Jacobi equation in cylindrical coordinates (ρ, ϕ, z), find the action variables, and express the Hamiltonian in terms of these. Find the frequencies $(\omega_\rho, \omega_\phi, \omega_z)$.
(c) Separate the Hamilton-Jacobi equation in spherical polar coordinates (r, θ, ϕ), find the action variables, and express the Hamiltonian in terms of these. Find the frequencies $(\omega_r, \omega_\theta, \omega_\phi)$.

Solution

(a) Writing down the Hamilton-Jacobi equation in cartesian coordinates and solving it by separation of variables proceeds as in Exercise 8.06. The momenta conjugate to x, y, and z are

$$p_x = \sqrt{2m\alpha_x - m^2\omega^2x^2}, \quad p_y = \sqrt{2m\alpha_y - m^2\omega^2y^2}, \quad p_z = \sqrt{2m\alpha_z - m^2\omega^2z^2},$$

where the separation constants α_x, α_y, and α_z are the energies associated with each degree of freedom, with $\alpha_x + \alpha_y + \alpha_z = E$. The action variables are then

$$\begin{aligned} I_x &= \frac{1}{2\pi}\oint\sqrt{2m\alpha_x - m^2\omega^2x^2}\,dx = \alpha_x/\omega, \\ I_y &= \frac{1}{2\pi}\oint\sqrt{2m\alpha_y - m^2\omega^2y^2}\,dy = \alpha_y/\omega, \\ I_z &= \frac{1}{2\pi}\oint\sqrt{2m\alpha_z - m^2\omega^2z^2}\,dz = \alpha_z/\omega. \end{aligned}$$

The total energy E, in terms of the action variables, is

$$E = \omega(I_x + I_y + I_z),$$

and the frequencies associated with the three degrees of freedom are

$$\omega_x = \frac{\partial E}{\partial I_x} = \omega, \quad \omega_y = \frac{\partial E}{\partial I_y} = \omega, \quad \omega_z = \frac{\partial E}{\partial I_z} = \omega.$$

(b) The Hamilton-Jacobi equation in cylindrical coordinates is

$$\frac{1}{2m}\left(\left(\frac{\partial W}{\partial \rho}\right)^2 + \frac{1}{\rho^2}\left(\frac{\partial W}{\partial \phi}\right)^2 + \left(\frac{\partial W}{\partial z}\right)^2\right) + \frac{1}{2}m\omega^2\left(\rho^2 + z^2\right) = E.$$

We try a solution of the form

$$W = R(\rho) + \Phi(\phi) + Z(z).$$

The Hamilton-Jacobi equation becomes

$$2m\rho^2\left[\frac{1}{2m}\left(\left(\frac{dR}{d\rho}\right)^2 + \left(\frac{dZ}{dz}\right)^2\right) + \frac{1}{2}m\omega^2\left(\rho^2 + z^2\right) - E\right] = -\left(\frac{d\Phi}{d\phi}\right)^2.$$

The left-hand side is a function only of ρ and z, whereas the right-hand side is a function only of ϕ. Both sides must equal a constant. We set

$$2m\rho^2\left[\frac{1}{2m}\left(\left(\frac{dR}{d\rho}\right)^2 + \left(\frac{dZ}{dz}\right)^2\right) + \frac{1}{2}m\omega^2\left(\rho^2 + z^2\right) - E\right] = -L_z^2, \qquad \left(\frac{d\Phi}{d\phi}\right)^2 = L_z^2,$$

where the separation constant L_z turns out to be the z-component of the angular momentum. The ρz-equation can be written

$$\left(\frac{1}{2m}\left(\frac{dR}{d\rho}\right)^2 + \frac{1}{2}m\omega^2\rho^2 + \frac{L_z^2}{2m\rho^2}\right) + \left(\frac{1}{2m}\left(\frac{dZ}{dz}\right)^2 + \frac{1}{2}m\omega^2 z^2\right) = E.$$

The first term on the left is a function only of ρ, the second term is a function only of z, and their sum is the constant E. Each term must equal a constant. We set

$$\frac{1}{2m}\left(\frac{dR}{d\rho}\right)^2 + \frac{1}{2}m\omega^2\rho^2 + \frac{L_z^2}{2m\rho^2} = E_\perp, \qquad \frac{1}{2m}\left(\frac{dZ}{dz}\right)^2 + \frac{1}{2}m\omega^2 z^2 = E_\parallel,$$

where $E_\perp + E_\parallel = E$. The separation constants $E_\perp$ and $E_\parallel$ are the energies associated with the "perpendicular" (or $\rho\phi$) and "parallel" (or z) motions. The variables are now completely separated, and we can read off the momenta conjugate to ρ, ϕ, and z,

$$p_\rho = \frac{\partial W}{\partial \rho} = \frac{dR}{d\rho} = \sqrt{2mE_\perp - m^2\omega^2\rho^2 - L_z^2/\rho^2},$$

$$p_\phi = \frac{\partial W}{\partial \phi} = \frac{d\Phi}{d\phi} = L_z,$$

$$p_z = \frac{\partial W}{\partial z} = \frac{dZ}{dz} = \sqrt{2mE_{\parallel} - m^2\omega^2 z^2}\,.$$

The action variable associated with the ρ degree of freedom is

$$\begin{aligned} I_\rho &= \frac{1}{\pi}\int_{\rho_{min}}^{\rho_{max}} \sqrt{2mE_\perp - m^2\omega^2\rho^2 - L_z^2/\rho^2}\, d\rho \\ &= \frac{1}{\pi}\int_{\rho_{min}}^{\rho_{max}} \left[\frac{mE_\perp - m^2\omega^2\rho^2}{\sqrt{2mE_\perp - m^2\omega^2\rho^2 - L_z^2/\rho^2}} + \frac{mE_\perp}{\sqrt{}} - \frac{L_z^2}{\rho^2\sqrt{}}\right] d\rho \end{aligned}$$

where the square root is taken positive and where ρ_{min} and ρ_{max} are the inner and outer turning radii (the values of ρ which make the square root zero). The first term is

$$\frac{1}{\pi}\int_{\rho_{min}}^{\rho_{max}} \frac{mE_\perp - m^2\omega^2\rho^2}{\sqrt{}}\, d\rho = \frac{1}{2\pi}\int_{\rho_{min}}^{\rho_{max}} d\left(\rho\sqrt{}\right) = 0,$$

since the square root is zero at the turning radii. The second term is

$$\begin{aligned} \frac{1}{\pi}\int_{\rho_{min}}^{\rho_{max}} \frac{mE_\perp}{\sqrt{}}\, d\rho &= \frac{E_\perp}{2\pi\omega}\int_{\rho_{min}}^{\rho_{max}} \frac{2\rho\, d\rho}{\sqrt{\dfrac{2E_\perp}{m\omega^2}\rho^2 - \rho^4 - \dfrac{L_z^2}{m^2\omega^2}}} \\ &= \frac{E_\perp}{2\pi\omega}\int_{\rho_{min}}^{\rho_{max}} \frac{2\rho\, d\rho}{\sqrt{\dfrac{E_\perp^2}{m^2\omega^4}\left(1 - \dfrac{\omega^2 L^2}{E_\perp^2}\right) - \left(\dfrac{E_\perp}{m\omega^2} - \rho^2\right)^2}} \end{aligned}$$

The integration can be performed by setting

$$\frac{E_\perp}{m\omega^2} - \rho^2 = \frac{E_\perp}{m\omega^2}\sqrt{1 - \frac{\omega^2 L_z^2}{E_\perp^2}}\cos\alpha, \qquad 2\rho\, d\rho = \frac{E_\perp}{m\omega^2}\sqrt{1 - \frac{\omega^2 L_z^2}{E_\perp^2}}\sin\alpha\, d\alpha.$$

As ρ increases from ρ_{min} to ρ_{max} with the square root positive, α increases from 0 to π and we must take $\sqrt{\sin^2\alpha} = \sin\alpha$. The second term then gives $E_\perp/2\omega$. The third term is

$$-\frac{1}{\pi}\int_{\rho_{min}}^{\rho_{max}}\frac{L_z^2}{\rho^2\sqrt{}}d\rho = -\frac{|L_z|}{2\pi}\int_{\rho_{min}}^{\rho_{max}}\frac{2d\rho/\rho^3}{\sqrt{\dfrac{2mE_\perp}{L_z^2}\dfrac{1}{\rho^2}-\dfrac{m^2\omega^2}{L_z^2}-\dfrac{1}{\rho^4}}}$$

$$= -\frac{|L_z|}{2\pi}\int_{\rho_{min}}^{\rho_{max}}\frac{2d\rho/\rho^3}{\sqrt{\dfrac{m^2E_\perp^2}{L_z^4}\left(1-\dfrac{\omega^2L_z^2}{E_\perp^2}\right)-\left(\dfrac{1}{\rho^2}-\dfrac{mE_\perp}{L_z^2}\right)^2}}$$

The integration can be performed by setting

$$\frac{1}{\rho^2}-\frac{mE_\perp}{L_z^2}=\frac{mE_\perp}{L_z^2}\sqrt{1-\frac{\omega^2L_z^2}{E_\perp^2}}\cos\beta, \qquad \frac{2d\rho}{\rho^3}=\frac{mE_\perp}{L_z^2}\sqrt{1-\frac{\omega^2L_z^2}{E_\perp^2}}\sin\beta\, d\beta.$$

As ρ increases from ρ_{min} to ρ_{max} with the square root positive, β increases from 0 to π and we must take $\sqrt{\sin^2\beta}=\sin\beta$. The third term then gives $-|L_z|/2$. Putting these results together, we find the action variable associated with the ρ degree of freedom

$$I_\rho = \frac{E_\perp}{2\omega}-\frac{|L_z|}{2}.$$

The action variable associated with the ϕ degree of freedom is

$$I_\phi = \frac{1}{2\pi}\oint L_z\, d\phi = |L_z|.$$

The action variable associated with the z degree of freedom is

$$I_z = \frac{1}{\pi}\int_{z_{min}}^{z_{max}}\sqrt{2mE_\parallel - m^2\omega^2z^2}\, dz = E_\parallel/\omega.$$

These results can be inverted to express the energy in terms of the action variables,

$$E = \omega(2I_\rho + I_\phi + I_z).$$

The frequencies associated with the three degrees of freedom are then

$$\omega_\rho = \frac{\partial E}{\partial I_\rho} = 2\omega, \qquad \omega_\phi = \frac{\partial E}{\partial I_\phi} = \omega, \qquad \omega_z = \frac{\partial E}{\partial I_z} = \omega.$$

(c) Writing down the Hamilton-Jacobi equation in spherical polar coordinates and solving it by separation of variables proceeds as in Exercise 8.06 and in *Lagrangian and Hamiltonian Mechanics*, pages 157, 158, 178, 179. The momenta conjugate to r, θ, and ϕ are

$$p_r = \sqrt{2mE - m^2\omega^2 r^2 - \frac{L^2}{r^2}}, \qquad p_\theta = \sqrt{L^2 - \frac{L_z^2}{\sin^2\theta}}, \qquad p_\phi = L_z,$$

where the separation constants are the total energy E, the magnitude of the angular momentum L, and the z-component of the angular momentum L_z. The action variable associated with the radial degree of freedom is

$$I_r = \frac{1}{\pi}\int_{r_{min}}^{r_{max}} \sqrt{2mE - m^2\omega^2 r^2 - L^2/r^2}\, dr.$$

This has the same form as the I_ρ integration in part (b), so we obtain

$$I_r = \frac{E}{2\omega} - \frac{L}{2}.$$

The action variables associated with the angular degrees of freedom are

$$I_\theta = \frac{1}{2\pi}\oint\sqrt{L^2 - \frac{L_z^2}{\sin^2\theta}}\, d\theta = L - |L_z|, \qquad I_\phi = \frac{1}{2\pi}\oint L_z\, d\phi = |L_z|.$$

The energy E, in terms of the action variables, is

$$E = \omega(2I_r + I_\theta + I_\phi).$$

The frequencies associated with the three degrees of freedom are then

$$\omega_r = \frac{\partial E}{\partial I_r} = 2\omega, \quad \omega_\theta = \frac{\partial E}{\partial I_\theta} = \omega, \quad \omega_\phi = \frac{\partial E}{\partial I_\phi} = \omega.$$

Exercise 9.06

A particle of mass m moves in a central potential

$$V = -\frac{k}{r} + \frac{h}{r^2}.$$

(a) Find the action variable I_r in terms of the energy E and total angular momentum L.
(b) Use your result to express the energy in terms of the action variables (I_r, I_θ, I_ϕ).
(c) Find the frequencies $(\omega_r, \omega_\theta, \omega_\phi)$. Under what conditions (on the action variables) is the motion periodic?

Solution

(a) The action variable I_r is given by

$$I_r = \frac{\sqrt{2m}}{2\pi}\oint\sqrt{E + \frac{k}{r} - \frac{h}{r^2} - \frac{L^2}{2mr^2}}\,dr.$$

We rewrite this

$$\frac{2\pi I_r}{\sqrt{2m}} = \oint\frac{(E + k/2r)\,dr}{\sqrt{E + k/r - (L^2 + 2mh)/2mr^2}} + \oint\frac{(k/2r)\,dr}{\sqrt{\;}} - \oint\frac{((L^2 + 2mh)/2mr^2)\,dr}{\sqrt{\;}}.$$

The integrals can be performed as in *Lagrangian and Hamiltonian Mechanics*, page 180. The first is zero since the integrand is the exact differential $d(r\sqrt{E + k/r - (L^2 + 2mh)/r^2})$. The second simplifies with the substitution

$$\frac{r}{a} = 1 - e\cos B, \quad \text{where} \quad a = \frac{k}{(-2E)} \quad \text{and} \quad e = \sqrt{1 + \frac{2(L^2 + 2mh)E}{mk^2}},$$

becoming

$$\frac{1}{\sqrt{2m}}\sqrt{\frac{mk^2}{(-2E)}}\oint dB = \frac{2\pi}{\sqrt{2m}}\sqrt{\frac{mk^2}{(-2E)}}.$$

The third simplifies with the substitution

$$\frac{a(1 - e^2)}{r} = 1 + e\cos A,$$

becoming

$$-\frac{\sqrt{L^2+2mh}}{\sqrt{2m}}\oint dA = -\frac{2\pi\sqrt{L^2+2mh}}{\sqrt{2m}}.$$

We thus find the action variable associated with the radial degree of freedom

$$I_r = \sqrt{\frac{mk^2}{(-2E)}} - \sqrt{L^2+2mh}.$$

(b) The action variables associated with the angular degrees of freedom are obtained in *Lagrangian and Hamiltonian Mechanics*, pages 178 and 179,

$$I_\theta = L - |L_z|, \qquad I_\phi = |L_z|.$$

These results can be inverted to express the energy in terms of the action variables,

$$E = -\frac{mk^2}{2\left(I_r + \sqrt{(I_\theta + I_\phi)^2 + 2mh}\right)^2}.$$

(c) The frequencies associated with the three degrees of freedom are then

$$\omega_r = \frac{\partial E}{\partial I_r} = \frac{mk^2}{\left(I_r + \sqrt{(I_\theta + I_\phi)^2 + 2mh}\right)^3}, \qquad \omega_{\theta,\phi} = \frac{\partial E}{\partial I_{\theta,\phi}} = \omega_r \frac{I_\theta + I_\phi}{\sqrt{(I_\theta + I_\phi)^2 + 2mh}}.$$

Periodic orbits occur when the frequency ratio

$$\frac{\omega_r}{\omega_{\theta,\phi}} = \sqrt{1 + \frac{2mh}{(I_\theta + I_\phi)^2}} = \alpha$$

is a rational number, $\alpha = n_r/n_{\theta,\phi}$ say. They are orbits for which

$$L^2 = (I_\theta + I_\phi)^2 = \frac{2mh}{(n_r/n_{\theta,\phi})^2 - 1}.$$

We have seen in Exercise 1.13 that in one cycle of r the angle in the plane of the orbit increases by $2\pi/\alpha$, so in n_r cycles it increases by $2\pi n_r/\alpha$. If this equals an integral multiple of 2π, say $2\pi n_{\theta,\phi}$, then the orbit closes and is periodic. This is another way to obtain the preceding condition, $\alpha = n_r/n_{\theta,\phi}$, for periodic orbits.

Exercise 9.07

A particle of mass m and charge e moves in a three-dimensional isotropic oscillator well $V = \frac{1}{2}m\omega^2 r^2$, on which is superimposed a uniform magnetic field $\mathbf{B}$. Choosing the symmetric gauge for the vector potential $\mathbf{A} = \frac{1}{2}\mathbf{B}\times\mathbf{r}$ and cylindrical coordinates (ρ, ϕ, z) with z-axis in the direction of the magnetic field, show that the time-independent Hamilton-Jacobi equation separates, obtain the action variables (I_ρ, I_ϕ, I_z), and express the Hamiltonian in terms of these (Ans. $H = (2I_\rho + I_\phi)\sqrt{\omega^2 + \omega_L^2} \mp I_\phi\omega_L + I_z\omega$, where $\omega_L = eB/2mc$ is the Larmor frequency)

Solution

Referring to Exercise 8.12, we see that the Hamilton-Jacobi equation is

$$\frac{1}{2m}\left[\left(\frac{\partial W}{\partial \rho}\right)^2 + \left(\frac{1}{\rho}\frac{\partial W}{\partial \phi} - \frac{eB}{2c}\rho\right)^2 + \left(\frac{\partial W}{\partial z}\right)^2\right] + \frac{1}{2}m\omega^2\left(\rho^2 + z^2\right) = E.$$

We try a solution of the form $W = R(\rho) + \Phi(\phi) + Z(z)$. The variables separate,

$$\frac{1}{2m}\left[\left(\frac{dR}{d\rho}\right)^2 + \left(\frac{L_z}{\rho} - \frac{eB}{2c}\rho\right)^2\right] + \frac{1}{2}m\omega^2\rho^2 = E_\perp,$$

$$\frac{1}{2m}\left(\frac{dZ}{dz}\right)^2 + \frac{1}{2}m\omega^2 z^2 = E_\parallel, \qquad \frac{d\Phi}{d\phi} = L_z,$$

where $E_\perp + E_\parallel = E$. The momenta conjugate to ρ, ϕ, and z are

$$p_\rho = \frac{\partial W}{\partial \rho} = \frac{dR}{d\rho} = \sqrt{2mE_\perp - \left(\frac{L_z}{\rho} - \frac{eB}{2c}\rho\right)^2 - m^2\omega^2\rho^2},$$

$$p_\phi = \frac{\partial W}{\partial \phi} = \frac{d\Phi}{d\phi} = L_z, \qquad p_z = \frac{\partial W}{\partial z} = \frac{dZ}{dz} = \sqrt{2mE_\parallel - m^2\omega^2 z^2}.$$

The action variable associated with the ρ degree of freedom is

$$I_\rho = \frac{1}{\pi}\int_{\rho_{min}}^{\rho_{max}} \sqrt{2mE_\perp - \left(\frac{L_z}{\rho} - \frac{eB}{2c}\rho\right)^2 - m^2\omega^2\rho^2}\, d\rho.$$

The quantity under the square root can be written

$$2m(E_\perp + \omega_L L_z) - L_z^2/\rho^2 - m^2(\omega_L^2 + \omega^2)\rho^2$$

where $\omega_L = eB/2mc$ is the Larmor frequency. Then, dividing and multiplying the integrand in I_ρ by the square root, we obtain

$$I_\rho = \frac{1}{\pi}\int_{\rho_{min}}^{\rho_{max}} \left[\frac{m(E_\perp + \omega_L L_z) - m^2(\omega_L^2 + \omega^2)\rho^2}{\sqrt{2m(E_\perp + \omega_L L_z) - L_z^2/\rho^2 - m^2(\omega_L^2 + \omega^2)\rho^2}} + \frac{m(E_\perp + \omega_L L_z) + m|L_z|\sqrt{\omega_L^2 + \omega^2}}{\sqrt{}} - \frac{L_z^2/\rho^2 + m|L_z|\sqrt{\omega_L^2 + \omega^2}}{\sqrt{}} \right] d\rho \, .$$

The first term in I_ρ is the differential $d\left(\frac{1}{2}\rho\sqrt{}\right)$ and gives zero contribution since the square root is zero at the turning points. The second term in I_ρ is

$$\frac{1}{\pi}\int_{\rho_{min}}^{\rho_{max}} \frac{m(E_\perp + \omega_L L_z) + m|L_z|\sqrt{\omega_L^2 + \omega^2}}{\sqrt{2m(E_\perp + \omega_L L_z)\rho^2 - L_z^2 - m^2(\omega_L^2 + \omega^2)\rho^4}}\rho d\rho$$

$$= \frac{1}{2\pi}\left(\frac{E_\perp + \omega_L L_z}{\sqrt{\omega_L^2 + \omega^2}} + |L_z|\right)\int_{u_{min}}^{u_{max}} \frac{du}{\sqrt{\dfrac{(E_\perp + \omega_L L_z)^2 - (\omega_L^2 + \omega^2)L_z^2}{m^2(\omega_L^2 + \omega^2)^2} - \left(u - \dfrac{E_\perp + \omega_L L_z}{m(\omega_L^2 + \omega^2)}\right)^2}}$$

where we have introduced the integration variable $u = \rho^2$. The u-integration gives π so the contribution of the second term to I_ρ is

$$\frac{1}{2}\left(\frac{E_\perp + \omega_L L_z}{\sqrt{\omega_L^2 + \omega^2}} + |L_z|\right).$$

To simplify the third term in I_ρ, we introduce the variable

$$\eta = -|L_z|/\rho + m\sqrt{\omega_L^2 + \omega^2}\,\rho, \quad d\eta = \left(|L_z|/\rho^2 + m\sqrt{\omega_L^2 + \omega^2}\right)d\rho,$$
$$\eta^2 = L_z^2/\rho^2 + m^2(\omega_L^2 + \omega^2)\rho^2 - 2m|L_z|\sqrt{\omega_L^2 + \omega^2}\,.$$

Note that η increases monotonely as ρ increases. The third term in I_ρ becomes

$$-\frac{|L_z|}{\pi}\int_{\eta_{min}}^{\eta_{max}} \frac{d\eta}{\sqrt{2m\left(E_\perp + \omega_L L_z - |L_z|\sqrt{\omega_L^2+\omega^2}\right) - \eta^2}}.$$

The η-integration gives π, so the contribution of the third term to I_ρ is $-|L_z|$. Putting together our results, we find the action variable I_ρ associated with the ρ degree of freedom,

$$I_\rho = \frac{1}{2}\left(\frac{E_\perp + \omega_L L_z}{\sqrt{\omega_L^2+\omega^2}} - |L_z|\right).$$

The action variables associated with the ϕ and z degrees of freedom are

$$I_\phi = \frac{1}{2\pi}\oint L_z\, d\phi = |L_z| \quad \text{and} \quad I_z = \frac{1}{\pi}\int_{z_{min}}^{z_{max}} \sqrt{2mE_\parallel - m^2\omega^2 z^2}\, dz = \frac{E_\parallel}{\omega}.$$

The total energy E can now be written in terms of the action variables,

$$E = (2I_\rho + I_\phi)\sqrt{\omega_L^2+\omega^2} \mp \omega_L I_\phi + \omega I_z.$$

The upper sign is to be used if L_z is positive, the lower if L_z is negative. Note that this expression for E reduces to the result of Exercise 9.05(c) if we switch off the magnetic field (set $\omega_L = 0$). The frequencies associated with the ρ, ϕ, and z degrees of freedom are

$$\omega_\rho = \frac{\partial E}{\partial I_\rho} = 2\sqrt{\omega_L^2+\omega^2}, \quad \omega_\phi = \frac{\partial E}{\partial I_\phi} = \sqrt{\omega_L^2+\omega^2} \mp \omega_L, \quad \omega_z = \frac{\partial E}{\partial I_z} = \omega.$$

Exercise 9.08

A simple harmonic oscillator with time-dependent frequency $\omega(t)$ has a Hamiltonian

$$H = \frac{p^2}{2m} + \frac{1}{2}m\omega^2(t)q^2.$$

(a) Transform from (q,p) variables to (instantaneous) action-angle variables (ϕ, I). Find, in particular, the Hamiltonian to be used with the action-angle variables.
(b) Write down Hamilton's equations of motion for the action-angle variables.

Solution

(a) We transform to instantaneous action-angle variables (ϕ, I) by setting

$$q = \sqrt{\frac{2I}{m\omega(t)}}\sin\phi, \qquad p = \sqrt{2m\omega(t)I}\cos\phi.$$

The Hamiltonian K for the action-angle variables is given by

$$K(\phi, I) = H(q,p) + \left(\frac{\partial W}{\partial t}\right)_{q,I}$$

where

$$W(q,I) = I(\phi + \sin\phi\cos\phi)$$

is the generating function of the transformation. The angle ϕ here is understood to be expressed in terms of q and I through the relation

$$\sin\phi = q\sqrt{m\omega/2I}.$$

The first term in K is

$$H(q,p) = \frac{1}{2m}\left(\sqrt{2m\omega I}\cos\phi\right)^2 + \frac{1}{2}m\omega^2\left(\sqrt{\frac{2I}{m\omega}}\sin\phi\right)^2 = \omega I.$$

The second is

$$\left(\frac{\partial W}{\partial t}\right)_{q,I} = 2I\cos^2\phi\left(\frac{\partial\phi}{\partial t}\right)_{q,I} = \frac{I\sin 2\phi}{2\omega}\frac{d\omega}{dt}.$$

The Hamiltonian for the action-angle variables is thus

$$K(\phi, I) = \omega I + \frac{I\sin 2\phi}{2\omega}\frac{d\omega}{dt}.$$

Note that it depends on the instantaneous angle variable as well as on the instantaneous action variable, and also that it depends linearly on the rate of change $d\omega/dt$ of the time-varying parameter $\omega(t)$.

(b) Hamilton's equations for the action-angle variables are

$$\frac{d\phi}{dt} = \frac{\partial K}{\partial I} = \omega + \frac{\sin 2\phi}{2\omega}\frac{d\omega}{dt}, \qquad \frac{dI}{dt} = -\frac{\partial K}{\partial\phi} = -I\frac{\cos 2\phi}{\omega}\frac{d\omega}{dt}.$$

Exercise 9.09

Consider again the simple plane pendulum undergoing small amplitude oscillations, and suppose that the length ℓ is shortened adiabatically, this time by pulling the string up through a small hole in the ceiling. Using elementary mechanics, show that the energy of oscillation E_{osc} increases such that $E_{osc}\sqrt{\ell}$ remains constant.

Solution

According to Newton's second law the tension f in the string, as the length ℓ of the string is changed, is given by

$$f = mg\cos\theta - m(\ddot{\ell} - \ell\dot{\theta}^2).$$

This tension equals the force we must exert on the string, so the work we do in changing the length of the string is $-\int f\,d\ell$. We are interested here in small amplitude oscillations, for which $\cos\theta \approx 1 - \frac{1}{2}\theta^2$. Also, we are interested in the adiabatic limit in which the pendulum makes many oscillations in the time the length changes by a small amount. We can then, to a good approximation, replace f in the expression for the work done by its average over an oscillation. Now the average of $mg\cos\theta$ is

$$\langle mg\cos\theta\rangle \approx mg - \tfrac{1}{2}mg\langle\theta^2\rangle = mg - \tfrac{1}{4}mg\theta_{max}^2 = mg - E_{osc}/2\ell$$

where $E_{osc} = \frac{1}{2}m(g/\ell)(\ell\theta_{max})^2$ is the energy of oscillation ($\sqrt{g/\ell}$ is the angular frequency and $\ell\theta_{max}$ is the amplitude of oscillation). The average of $-m\ddot{\ell}$ is zero, and the average of $m\ell\dot{\theta}^2$ is

$$\langle m\ell\dot{\theta}^2\rangle = \tfrac{1}{2}m\ell\dot{\theta}_{max}^2 = E_{osc}/\ell$$

where $E_{osc} = \frac{1}{2}m(\ell\dot{\theta}_{max})^2$ is again the energy of oscillation. The tension in the string, averaged over an oscillation, is thus

$$\langle f\rangle \approx mg - E_{osc}/2\ell + E_{osc}/\ell = mg + E_{osc}/2\ell.$$

The work done on the system as the length of the string is changed adiabatically by a small amount $\delta\ell$ is $-\langle f\rangle\delta\ell$. This equals the change in the total energy of the system, which consists of a change $-mg\delta\ell$ in the equilibrium energy plus a change δE_{osc} in the oscillation energy. We have

$$-mg\delta\ell - (E_{osc}/2\ell)\delta\ell = -mg\delta\ell + \delta E_{osc},$$

which gives

$$\frac{dE_{osc}}{E_{osc}} = -\frac{d\ell}{2\ell}.$$

Integrating this, we find $\ln E_{osc} = -\ln\sqrt{\ell} + \text{const.}$, so as the length of the string is changed adiabatically, the energy of oscillation changes such that $E_{osc}\sqrt{\ell}$ remains constant. We have thus shown that this quantity, proportional to the action variable $I = E_{osc}\sqrt{\ell/g}$, is an adiabatic invariant.

Exercise 9.10

A particle of mass m moves in one dimension x between rigid walls at $x = 0$ and at $x = \ell$. Using elementary mechanics:
(a) Show that the average (outward) force on the walls is $F = 2E/\ell$ where E is the (kinetic) energy of the particle.
(b) Suppose now that the wall at $x = \ell$ is moved adiabatically. The energy of the particle then changes as a result of its collisions with the moving wall. Show that $\delta E = -(2E/\ell)\delta\ell$.
(c) Hence show that $E\ell^2$ remains constant under this adiabatic change. Compare this result with that given by "invariance of the action variable."

Solution

(a) A particle of mass m moves back and forth between rigid walls at $x = 0$ and at $x = \ell$. If the (kinetic) energy of the particle is E, its speed is $v = \sqrt{2E/m}$. The number of collisions the particle makes with one of the walls per second is $v/2\ell$, and the (outward) momentum the particle transfers to the wall per collision is $2mv$. The average (outward) force the particle exerts on each of the walls is thus

$$F = 2mv \times v/2\ell = 2E/\ell.$$

(b) Now suppose that the wall at $x = \ell$ is moved adiabatically through a small displacement $\delta\ell$. The work done on the system is $\delta W = -F\delta\ell$, and this equals the change δE in energy of the system,

$$\delta E = -(2E/\ell)\delta\ell.$$

Another way to obtain this result is to observe that the speed of the particle is changed as a result of its collision with the moving wall. If the speed of the particle before collision is v and the velocity of the wall is u, the speed of the particle after collision is $v - 2u$ (to see this, analyze the collision in the rest frame of the wall). The change in the kinetic energy of the particle in the collision is thus

$$\tfrac{1}{2}m(v-2u)^2 - \tfrac{1}{2}mv^2 \approx -2mvu,$$

where we have assumed that the velocity u of the wall is small and have hence dropped the term quadratic in u. The number of such collisions in time δt is $(v/2\ell)\delta t$ so the change in the energy of the particle in this time is

$$\delta E = -2mvu \times (v/2\ell)\delta t = -(2E/\ell)\delta\ell$$

where we have set $\delta\ell = u\delta t$. This agrees with the result obtained from using the work-energy theorem.

(c) Integrating the result obtained in part (b),

$$\frac{dE}{E} = -\frac{2d\ell}{\ell},$$

we see that $\ln E = -\ln \ell^2 + \text{const.}$, and thus the quantity $E\ell^2$ remains constant in an adiabatic change in the size of the box. Now the action variable for this system is

$$I = \frac{1}{2\pi} mv \times 2\ell = \frac{1}{\pi}\sqrt{2mE}\,\ell,$$

so invariance of $E\ell^2$ under adiabatic expansion is the same as invariance of I.

Exercise 9.11

Consider again the Hannay hoop. Write down the Lagrangian (the kinetic energy of the bead in an inertial frame) using as generalized coordinate the displacement s of the bead around the hoop from some fixed point on the hoop. Assume that the hoop is rotating with angular velocity $\Omega = d\theta/dt$. Find the Hamiltonian, and write down Hamilton's equations of motion. Average $(1/\ell)\int_0^\ell \cdots ds$ the right-hand side of these over the position of the bead around the hoop, and integrate with respect to time to obtain the Hannay displacement $\langle \Delta s \rangle = -(2A/\ell)\Delta\theta$ in the position of the bead. Here A is the total area enclosed by the hoop and $\Delta\theta$ the angle through which the hoop is turned. (Hint: the Hamiltonian is $H = \dfrac{p^2}{2m} - pr\sin\alpha\,\Omega - \dfrac{1}{2}mr^2\cos^2\alpha\,\Omega^2$)

Solution

The velocity of the bead, with respect to the hoop, is $\dot{s}$ tangential to the hoop, and the velocity of the point of the hoop at which the bead is instantaneously located, with respect to the inertial frame about which the hoop is rotating, is $r\Omega$ perpendicular to the radius vector. The angle between these velocities is $\pi/2 - \alpha$ so, using the cosine law to add them, we find that the speed v of the bead, with respect to the inertial frame, is given by

$$v^2 = \dot{s}^2 + r^2\Omega^2 + 2\dot{s}r\Omega\sin\alpha.$$

The kinetic energy of the bead is $\frac{1}{2}mv^2$, and since there is no potential energy this equals the Lagrangian,

$$L(s,\dot{s},t) = \tfrac{1}{2}m(\dot{s}^2 + r^2\Omega^2 + 2\dot{s}r\Omega\sin\alpha).$$

The momentum conjugate to s is

$$p = \frac{\partial L}{\partial \dot{s}} = m\dot{s} + mr\Omega\sin\alpha,$$

and the Hamiltonian is

$$H = p\dot{s} - L = \frac{1}{2}m\dot{s}^2 - \frac{1}{2}mr^2\Omega^2 = \frac{1}{2m}(p - mr\Omega\sin\alpha)^2 - \frac{1}{2}mr^2\Omega^2.$$

Hamilton's equations are then

$$\frac{ds}{dt} = \frac{\partial H}{\partial p} = \frac{p}{m} - r\Omega\sin\alpha, \qquad \frac{dp}{dt} = -\frac{\partial H}{\partial s}.$$

Under the conditions being considered here, the bead makes many circuits as the hoop is turned through, say, one rotation. We can then, to a good approximation, replace the right-hand sides of Hamilton's equations by their averages over the position of the bead around the hoop. This gives

$$\frac{ds}{dt} \approx \left\langle\frac{p}{m}\right\rangle - \frac{\Omega}{\ell}\int_0^\ell r\sin\alpha\, ds = \left\langle\frac{p}{m}\right\rangle - \frac{2A\Omega}{\ell}, \qquad \frac{dp}{dt} \approx -\frac{1}{\ell}\int_0^\ell \frac{\partial H}{\partial s} ds = 0,$$

where $A = \frac{1}{2}\int_0^\ell r\sin\alpha\, ds$ is the area enclosed by the hoop. Integrating with respect to time, we see that the conjugate momentum p is approximately constant, $\Delta p \approx 0$, and the change is s is

$$\Delta s = \int_0^T \frac{ds}{dt} dt = \frac{p}{m}T - \frac{2A}{\ell}\Delta\theta$$

where $\Delta\theta = \int_0^T \Omega\, dt$ is the angle through which the hoop is turned. The first term in Δs is the dynamical change, and the second term, $-(2A/\ell)\Delta\theta$, is the Hannay change.

Exercise 9.12

Consider the "generalized simple harmonic oscillator" with Hamiltonian

$$H = \tfrac{1}{2}(Xq^2 + 2Yqp + Zp^2)$$

where (X, Y, Z) are parameters with $XZ > Y^2$.
(a) Show that the trajectories in phase space are ellipses and hence find the action variable, showing that it is $I = E/\omega$ where $\omega = \sqrt{XZ - Y^2}$ is the frequency.
(b) Express the variables (q, p) in terms of the action-angle variables (ϕ, I). (There are various ways to do this; one way is to solve the Hamilton-Jacobi equation to find the appropriate generating function.)
(c) Suppose that the parameters $\mathbf{R} = (X, Y, Z)$ are changed adiabatically (but always with $XZ > Y^2$) so as to take the system around a closed circuit in parameter space. Show that the resulting Hannay angle is

$$\Delta\phi_H = \iint \frac{\mathbf{R}\cdot d\mathbf{S}}{4\omega^3}.$$

(J. H. Hannay, "Angle variable holonomy in adiabatic excursion of an integrable Hamiltonian," J. Phys. A **18**, 221-230 (1985); M. V. Berry, "Classical adiabatic angles and quantal adiabatic phase," J. Phys. A **18**, 15-27 (1985).)

Solution

If we let $\xi = \begin{bmatrix} q \\ p \end{bmatrix}$ denote a column vector of the canonical variables, the Hamiltonian can be written

$$H = \xi^T M \xi$$

where $\xi^T = \begin{bmatrix} q & p \end{bmatrix}$ is the transpose of ξ, and M is the symmetric matrix

$$M = \frac{1}{2}\begin{bmatrix} X & Y \\ Y & Z \end{bmatrix}.$$

Suppose we introduce new canonical variables ξ' by setting

$$\xi' = R\xi, \quad \xi = R^T\xi' \quad \text{where} \quad R = \begin{bmatrix} \cos\alpha & \sin\alpha \\ -\sin\alpha & \cos\alpha \end{bmatrix}.$$

This is a rotation of phase space through an angle α. The Hamiltonian for the new variables is

$$H' = \left(\xi'^T R\right) M \left(R^T \xi'\right) = \xi'^T \left(RMR^T\right) \xi' = \xi'^T M' \xi'$$

where $M' = RMR^T$. We can choose R so that the new matrix M' is diagonal,

$$M' = \frac{1}{2}\begin{bmatrix} \lambda_+ & 0 \\ 0 & \lambda_- \end{bmatrix}$$

where $\lambda_+/2$ and $\lambda_-/2$ are the (real) eigenvalues. Since the properties "determinant" and "trace" of a matrix are invariant under such transformations, we have

$$\lambda_+\lambda_- = XZ - Y^2 \quad \text{and} \quad \lambda_+ + \lambda_- = X + Y.$$

The new Hamiltonian becomes

$$H' = \frac{1}{2}\lambda_+ q'^2 + \frac{1}{2}\lambda_- p'^2.$$

Now if $XZ - Y^2 > 0$, the signs of λ_+ and λ_- are the same. The trajectories $H' = E$ in (q', p') phase space are then ellipses with axes along the q' and p' axes. The semi-q'-axis is $\sqrt{2E/\lambda_+}$, the semi-p'-axis is $\sqrt{2E/\lambda_-}$, and the ellipse encloses an area

$$\text{Area} = \pi\sqrt{2E/\lambda_+} \times \sqrt{2E/\lambda_-} = 2\pi E\Big/\sqrt{\lambda_+\lambda_-} = 2\pi E\Big/\sqrt{XZ - Y^2}.$$

The action variable is thus

$$I = \frac{\text{Area}}{2\pi} = \frac{E}{\sqrt{XZ - Y^2}},$$

and the angular frequency is

$$\omega = \frac{\partial E}{\partial I} = \sqrt{XZ - Y^2}.$$

Another way to approach this problem is to solve the Hamilton-Jacobi equation

$$\frac{1}{2}\left(Xq^2 + 2Yq\frac{dW}{dq} + Z\left(\frac{dW}{dq}\right)^2\right) = E$$

to obtain the generating function of the canonical transformation from the original variables (q, p) to action-angle variables (ϕ, I). The Hamilton-Jacobi equation gives

$$\frac{dW}{dq} = -\frac{Y}{Z}q + \frac{1}{Z}\sqrt{2ZE - (XZ - Y^2)q^2}.$$

Integration then gives

$$W = -\frac{Y}{2Z}q^2 + \frac{1}{Z}\int\sqrt{2ZE - (XZ - Y^2)q^2}\,dq.$$

The remaining integration can be performed by setting

$$q = \sqrt{\frac{2ZE}{XZ - Y^2}}\sin\phi, \qquad dq = \sqrt{\frac{2ZE}{XZ - Y^2}}\cos\phi\, d\phi.$$

We have

$$W = -\frac{Y}{2Z}q^2 + \frac{2E}{\sqrt{XZ - Y^2}}\int\cos^2\phi\, d\phi = -\frac{Y}{2Z}q^2 + \frac{E}{\sqrt{XZ - Y^2}}(\phi + \sin\phi\cos\phi).$$

As q is taken around one cycle, the variable ϕ increases by 2π and W increases by $2\pi E/\sqrt{XZ - Y^2}$. The action variable is then

$$I = \frac{\Delta W}{2\pi} = \frac{E}{\sqrt{XZ - Y^2}} = \frac{E}{\omega}$$

where we have introduced the angular frequency $\omega = \dfrac{\partial E}{\partial I} = \sqrt{XZ - Y^2}$. These results are the same as we obtained previously, but we can now continue on to find the corresponding angle variable. We replace, in W, the energy E by its expression in terms of the action variable I, obtaining

$$W(q, I) = -\frac{Y}{2Z}q^2 + I(\phi + \sin\phi\cos\phi) \quad \text{where} \quad \sin\phi = q\sqrt{\frac{\omega}{2ZI}}.$$

The resulting function $W(q, I)$ is the generating function of a canonical transformation from (q, p) variables to action-angle variables. The old momentum is

$$p = \left(\frac{\partial W}{\partial q}\right)_I = -\frac{Y}{Z}q + \sqrt{\frac{2\omega I}{Z}}\cos\phi.$$

The angle variable is

$$\left(\frac{\partial W}{\partial I}\right)_q = \phi;$$

that is, the angle variable is simply the variable ϕ which we introduced previously for purposes of integration. The transformation from the original (q, p) variables to action-angle variables (ϕ, I) is thus

$$q = \sqrt{\frac{2ZI}{\omega}}\sin\phi, \qquad p = \sqrt{\frac{2ZI}{\omega}}\left(-\frac{Y}{Z}\sin\phi + \frac{\omega}{Z}\cos\phi\right).$$

We now suppose that the parameters $\mathbf{R} = (X, Y, Z)$ are varied adiabatically so as to take the system around a closed circuit in the three-dimensional parameter space. We wish to find the resulting Hannay angle. For this we need the "flux density"

$$\mathcal{B} = \langle \nabla q \times \nabla p \rangle$$

where $\langle\ \rangle$ denotes angle averaging and where

$$\nabla q = \sqrt{2I}\sin\phi\nabla\left(\sqrt{\frac{Z}{\omega}}\right), \qquad \nabla p = -\sqrt{2I}\sin\phi\nabla\left(\sqrt{\frac{Z}{\omega}}\frac{Y}{Z}\right) + \sqrt{2I}\cos\phi\nabla\left(\sqrt{\frac{Z}{\omega}}\frac{\omega}{Z}\right).$$

The angle average of $\sin\phi\cos\phi$ is zero and that of $\sin^2\phi$ is 1/2, so we need only consider

$$\mathcal{B} = -2I\cdot\frac{1}{2}\cdot\nabla\left(\sqrt{\frac{Z}{\omega}}\right)\times\nabla\left(\sqrt{\frac{Z}{\omega}}\frac{Y}{Z}\right) = -\frac{I}{2}\nabla\left(\frac{Z}{\omega}\right)\times\nabla\left(\frac{Y}{Z}\right).$$

We have

$$\begin{aligned}\nabla\left(\frac{Z}{\omega}\right) &= \frac{1}{\omega}\nabla Z - \frac{Z}{2\omega^3}(X\nabla Z + Z\nabla X - 2Y\nabla Y)\\ &= \frac{1}{2\omega^3}\left(-Z^2\nabla X + 2YZ\nabla Y + (XZ - 2Y^2)\nabla Z\right)\end{aligned} \qquad \text{and} \qquad \nabla\left(\frac{Y}{Z}\right) = \frac{1}{Z}\nabla Y - \frac{Y}{Z^2}\nabla Z$$

so the expression for the "flux density" becomes

$$\mathcal{B} = -\frac{I}{4\omega^3}\begin{vmatrix}\nabla X & \nabla Y & \nabla Z\\ -Z^2 & 2YZ & XZ - 2Y^2\\ 0 & 1/Z & -Y/Z^2\end{vmatrix} = \frac{I}{4\omega^3}(X\nabla X + Y\nabla Y + Z\nabla Z) = \frac{I\,\mathbf{R}}{4\omega^3}.$$

The Hannay angle is then given by

$$\Delta\phi_H = \frac{\partial}{\partial I}\iint \mathcal{B}\cdot d\mathbf{S} = \iint \frac{\mathbf{R}\cdot d\mathbf{S}}{4\omega^3}.$$

CHAPTER X

NON-INTEGRABLE SYSTEMS

Exercise 10.01

Investigate the surface of section ($y = 0$, $p_y > 0$, E fixed) for the two-dimensional oscillator with Hamiltonian (refer to Exercise 9.02)

$$H = \frac{p_x^2}{2m} + \frac{p_y^2}{2m} + \frac{1}{2}m\omega_x^2 x^2 + \frac{1}{2}m\omega_y^2 y^2.$$

In particular, examine the nature of the sequences of points resulting from various starts,
(a) if ω_x/ω_y is an irrational number;
(b) if ω_x/ω_y is a rational number, $\omega_x/\omega_y = r/s$.

Solution

According to Exercise 9.02 the cartesian variables $(x, y; p_x, p_y)$ are related to the action-angle variables $(\phi_x, \phi_y; I_x, I_y)$ by the canonical transformation

$$x = \sqrt{\frac{2I_x}{m\omega_x}}\sin\phi_x, \qquad p_x = \sqrt{2m\omega_x I_x}\cos\phi_x,$$
$$y = \sqrt{\frac{2I_y}{m\omega_y}}\sin\phi_y, \qquad p_y = \sqrt{2m\omega_y I_y}\cos\phi_y.$$

The Hamiltonian, in terms of the action-angle variables, is

$$H = \omega_x I_x + \omega_y I_y.$$

Hamilton's equations show that the action variables I_x and I_y are constant in time and that the angle variables ϕ_x and ϕ_y increase uniformly with time,

$$\phi_x(t) = \phi_x(0) + \omega_x t,$$
$$\phi_y(t) = \phi_y(0) + \omega_y t.$$

Let us take a surface of section $y = 0$ with $p_y > 0$. Coordinates on the surface are then (x, p_x). Suppose we start the system at $t = 0$ on the surface of section at (x, p_x) and with energy E. The coordinate y is zero and its conjugate momentum is

$$p_y = \sqrt{2mE - p_x^2 - m^2\omega_x^2 x^2}.$$

The various starts for given energy E must lie within the ellipse

$$p_x^2 + m^2\omega_x^2 x^2 < 2mE.$$

The angle variable ϕ_y is initially zero. As time goes by, the angle variable ϕ_y increases, and at time $2\pi/\omega_y$ it equals 2π and the system is back at the surface of section. The angle variable ϕ_x is then

$$\phi_x(1) = \phi_x(0) + 2\pi(\omega_x/\omega_y),$$

and the coordinates of the point on the surface of section are

$$x(1) = \sqrt{\frac{2I_x}{m\omega_x}}\sin\left(\phi_x(0) + 2\pi(\omega_x/\omega_y)\right), \quad p_x(1) = \sqrt{2m\omega_x I_x}\cos\left(\phi_x(0) + 2\pi(\omega_x/\omega_y)\right).$$

This is repeated over and over, and we thus get a sequence of points

$$x(n) = \sqrt{\frac{2I_x}{m\omega_x}}\sin\left(\phi_x(0) + 2\pi n(\omega_x/\omega_y)\right), \quad p_x(n) = \sqrt{2m\omega_x I_x}\cos\left(\phi_x(0) + 2\pi n(\omega_x/\omega_y)\right),$$

on the surface of section. These points lie on the ellipse

$$\frac{1}{2m}p_x^2(n) + \frac{1}{2}m\omega_x^2 x^2(n) = \omega_x I_x.$$

If the frequency ratio ω_x/ω_y is an irrational number, the sequence never repeats and the points eventually cover the ellipse densely. If, on the other hand, the frequency ratio is a rational number, say $\omega_x/\omega_y = r/s$ where r and s are integers, then after s passages through the surface of section the angle variable ϕ_x increases by $2\pi r$ and the system is back at its start. We have a finite cycle of s points. For the two-dimensional oscillator these features do not depend on the start.

Exercise 10.02

A particle of mass m moves in a (two-dimensional) central force with potential

$$V = -\frac{k}{r} + \frac{h}{r^2}.$$

Using a computer or otherwise, plot the sequences of points (x, p_x) in the surface of section ($y = 0$, $p_y > 0$, E fixed) which result from representative starts (refer to Exercises 1.13 and 9.06).

Solution

According to Exercise 1.13 the equation of the orbit is

$$\frac{a(1-e^2)}{r} = 1 + e\cos\alpha(\theta - \theta_0)$$

where $\alpha = \sqrt{1+2mh/L^2}$. Here r and θ are polar coordinates in the plane of the orbit with θ_0 being the direction of a pericenter. The parameters a and e are the "semi-major-axis" and "eccentricity." They are related to the energy E and angular momentum L by

$$a = \frac{k}{(-2E)} \quad \text{and} \quad e = \sqrt{1+\frac{2E\alpha^2L^2}{mk^2}} = \sqrt{1-\frac{\alpha^2L^2}{mka}}.$$

The momentum p_r conjugate to r is given by the energy equation

$$\frac{p_r^2}{2m} + \frac{L^2}{2mr^2} - \frac{k}{r} + \frac{h}{r^2} = E.$$

Solving for p_r, we find

$$p_r = \sqrt{2mE + \frac{2mk}{r} - \frac{\alpha^2L^2}{r^2}} = \sqrt{\frac{mk}{a}}\sqrt{-1+\frac{2a}{r} - \frac{a^2(1-e^2)}{r^2}},$$

and substituting for r from the orbit equation, we find

$$p_r = \pm\sqrt{\frac{mk}{a}}\frac{e}{\sqrt{1-e^2}}\sin\alpha(\theta-\theta_0)$$

where we take the upper sign if L is positive and the lower sign if L is negative. The conjugate momentum p_r is the linear momentum in the r-direction. The momentum conjugate to θ is the angular momentum L. The *linear* momentum in the θ-direction is L/r.

Our surface of section is specified in terms of cartesian variables. These variables are obtained from the polar variables by the transformation equations

$$x = r\cos\theta, \qquad p_x = p_r\cos\theta - (L/r)\sin\theta,$$
$$y = r\sin\theta, \qquad p_y = p_r\sin\theta + (L/r)\cos\theta.$$

The surface of section is defined by

$$y = 0 \quad \text{with} \quad p_y > 0.$$

These conditions imply that on the surface of section

$$\theta = 0, 2\pi, 4\pi, \cdots \text{ for } L > 0 \quad \text{or} \quad \theta = -\pi, -3\pi, -5\pi, \cdots \text{ for } L < 0.$$

Coordinates on the surface of section are then ($x = \pm r$, $p_x = \pm p_r$) where again we take the upper sign if L is positive and the lower sign if L is negative. Sequences of points on the surface of section are given by

$$\frac{a(1-e^2)}{x(n)} = \pm(1 + e\cos 2\pi\alpha n), \qquad p_x(n) = \pm\sqrt{\frac{mk}{a}}\frac{e}{\sqrt{1-e^2}}\sin 2\pi\alpha n,$$

where $n = 0, 1, 2, \cdots$ and we have taken the start, for either sign of L, at pericenter. That is, we have taken $\theta_0 = 0$ for L positive and $\theta_0 = -\pi$ for L negative. The energy E and hence the "semi-major-axis" a on a given surface of section is constant. We can thus measure x in units of a and p_x in units of $\sqrt{mk/a}$, and the sequences of points become

$$\frac{1-e^2}{x(n)} = \pm(1 + e\cos 2\pi\alpha n), \qquad p_x(n) = \pm\frac{e}{\sqrt{1-e^2}}\sin 2\pi\alpha n.$$

The parameter α is best expressed here as

$$\frac{1}{\alpha} = \sqrt{1 - \frac{2\eta}{1-e^2}}$$

where $\eta = h/ka$ is a dimensionless measure of the strength of the $1/r^2$ term in the potential (see Exercise 1.13). We pick an eccentricity e and let $n = 0, 1, 2, \cdots$ to obtain one sequence of points, change e and obtain another sequence, etc. The result is Fig. 1, in which we have taken $\eta = 0.05$.

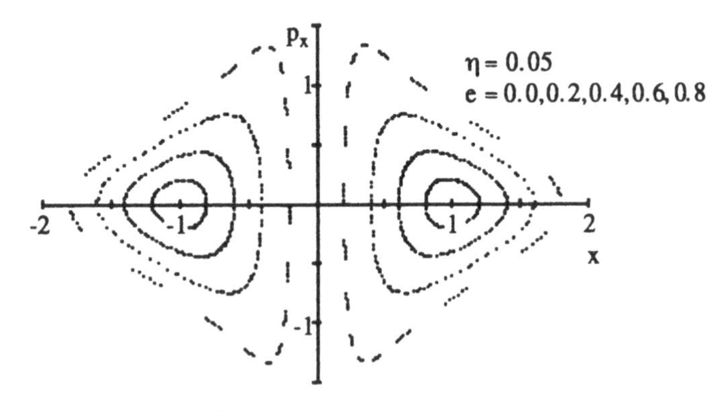

Ex. 10.02, Fig. 1

Eccentricity $e = 0$ gives the elliptic fixed points ($\pm 1, 0$), and eccentricities $e = 0.2$, 0.4, 0.6, and 0.8 give the successive surrounding "curves." The $x > 0$ set is for

$L > 0$, and the $x < 0$ set is for $L < 0$. Note that the outermost "curve" is almost a finite cycle of 17 points, so this is almost a periodic orbit. The reason is that (for these values of η and e) $17\alpha = 20.003845$ which is almost an integer ($\eta = 0.04995$ and $e = 0.8$ give $\alpha = 20/17$ exactly).

Exercise 10.03

Consider a system with Hamiltonian

$$H = \frac{p^2}{2m} + \frac{1}{2}k_0x^2 + \frac{1}{2}k_1x^2.$$

This is, of course, a simple harmonic oscillator with spring constant $k_0 + k_1$ and is exactly soluble. Suppose, however, that we regard the term $\frac{1}{2}k_1x^2$ as a perturbation. Find, to first order in the perturbation,
(a) the canonical transformation from the unperturbed action-angle variables to the perturbed action-angle variables;
(Ans.: the generator is $G = (k_1I'/4k_0)\sin 2\phi$)
(b) the Hamiltonian for the perturbed action-angle variables. Hence find the first order correction to the frequency of oscillation. Compare with the exact result.

Solution

(a) We write the Hamiltonian as

$$H = H_0 + H_1$$

where

$$H_0 = \frac{p^2}{2m} + \frac{1}{2}k_0x^2$$

is the unperturbed Hamiltonian and

$$H_1 = \frac{1}{2}k_1x^2$$

is the perturbation. The unperturbed system is a simple harmonic oscillator for which we can introduce action-angle variables (ϕ, I) by setting

$$x = \sqrt{\frac{2I}{m\omega_0}}\sin\phi, \qquad p = \sqrt{2m\omega_0 I}\cos\phi,$$

where $\omega_0 = \sqrt{k_0/m}$ is the unperturbed frequency. In terms of these action-angle variables the unperturbed Hamiltonian is

$$H_0(I) = \omega_0 I$$

and the perturbation is

$$H_1(\phi, I) = \frac{k_1 I}{m\omega_0}\sin^2\phi.$$

We wish to introduce new action-angle variables (ϕ', I') such that the new Hamiltonian is a function of the new action variable I' alone, $H'(I')$. To find the generator of the appropriate canonical transformation, we expand the perturbation,

$$H_1(\phi, I) = \frac{k_1 I}{4m\omega_0}\left(2 - e^{2i\phi} - e^{-2i\phi}\right),$$

so the Fourier components of H_1 are

$$h_0 = \frac{k_1 I}{2m\omega_0}, \qquad h_2 = -\frac{k_1 I}{4m\omega_0}, \qquad h_{-2} = -\frac{k_1 I}{4m\omega_0}.$$

The Fourier components of the appropriate generator are then (*Lagrangian and Hamiltonian Mechanics*, page 200)

$$g_2 = \frac{ih_2}{2\omega_0} = -\frac{ik_1 I}{8k_0}, \qquad g_{-2} = \frac{ih_{-2}}{(-2)\omega_0} = \frac{ik_1 I}{8k_0},$$

so the generator is

$$G(\phi, I) = -\frac{ik_1 I}{8k_0}\left(e^{2i\phi} - e^{-2i\phi}\right) = \frac{k_1 I}{4k_0}\sin 2\phi.$$

The canonical transformation generated by G is

$$\phi' = \phi + \frac{\partial G}{\partial I} = \phi + \frac{k_1}{4k_0}\sin 2\phi, \qquad I' = I - \frac{\partial G}{\partial \phi} = I - \frac{k_1 I}{2k_0}\cos 2\phi.$$

(b) The new Hamiltonian is, to first order in the perturbation,

$$H'(I') = H_0(I') + \langle H_1(\phi, I')\rangle$$

where $\langle H_1 \rangle$ is the average of the perturbation over a cycle. For this we can either take h_0 from part (a) or proceed directly,

$$\langle H_1(\phi,I')\rangle = \frac{k_1 I'}{m\omega_0}\langle \sin^2\phi\rangle = \frac{k_1 I'}{2m\omega_0}.$$

The new Hamiltonian is

$$H'(I') = \omega_0 I' + \frac{k_1 I'}{2m\omega_0}.$$

Another way to find the new Hamiltonian is simply to transform the old Hamiltonian, since from part (a) we know the appropriate canonical transformation. This gives

$$H'(\phi',I') = \omega_0\left(I' + \frac{k_1 I'}{2k_0}\cos 2\phi'\right) + \frac{k_1 I'}{m\omega_0}\sin^2\phi' = \omega_0 I'\left(1 + \frac{k_1}{2k_0}\right)$$

in agreement with the preceding approach.

The perturbed frequency of oscillation is

$$\omega' = \frac{\partial H'}{\partial I'} = \omega_0 + \frac{k_1}{2m\omega_0} = \omega_0\left(1 + \frac{k_1}{2k_0}\right),$$

which can be compared with the exact result

$$\omega = \sqrt{\frac{k_0 + k_1}{m}} = \omega_0\sqrt{1 + \frac{k_1}{k_0}} \approx \omega_0\left(1 + \frac{k_1}{2k_0} + \cdots\right).$$

Exercise 10.04

Find the continued fraction expansion of the following numbers, and write down the first five or so continued fraction approximates. Verify that these are closer to the number than (denominator)$^{-2}$. (a) $157/225$; (b) $\sqrt{2}$; (c) the golden ratio, $\gamma = (\sqrt{5}-1)/2$; (d) the base of natural logarithms, e.

Solution

(a) We have the continued fraction expansion

$$\frac{157}{225} = \cfrac{1}{1 + \cfrac{1}{2 + \cfrac{1}{3 + \cfrac{1}{4 + \cfrac{1}{5}}}}}$$

and continued fraction approximations

$$\frac{1}{1}=1,\quad \frac{1}{1+\frac{1}{2}}=\frac{2}{3},\quad \frac{1}{1+\frac{1}{2+\frac{1}{3}}}=\frac{7}{10},\quad \frac{1}{1+\frac{1}{2+\frac{1}{3+\frac{1}{4}}}}=\frac{30}{43},\quad \frac{1}{1+\frac{1}{2+\frac{1}{3+\frac{1}{4+\frac{1}{5}}}}}=\frac{157}{225}.$$

These differ from 157/225 by

$$\left|\frac{157}{225}-\frac{1}{1}\right|=\frac{68}{225}<\frac{1}{1^2},\quad \left|\frac{157}{225}-\frac{2}{3}\right|=\frac{7}{225}<\frac{1}{3^2}=\frac{1}{9},\quad \left|\frac{157}{225}-\frac{7}{10}\right|=\frac{1}{450}<\frac{1}{10^2}=\frac{1}{100},$$
$$\left|\frac{157}{225}-\frac{30}{43}\right|=\frac{1}{9675}<\frac{1}{43^2}=\frac{1}{1849},\quad \left|\frac{157}{225}-\frac{157}{225}\right|=0<\frac{1}{225^2}.$$

(b) We have the continued fraction expansion

$$\sqrt{2}=1+\frac{1}{2+\frac{1}{2+\frac{1}{2+\frac{1}{2+\frac{1}{2+\cdots}}}}}$$

and continued fraction approximates

$$1+\frac{1}{2}=\frac{3}{2},\quad 1+\frac{1}{2+\frac{1}{2}}=\frac{7}{5},\quad 1+\frac{1}{2+\frac{1}{2+\frac{1}{2}}}=\frac{17}{12},\quad 1+\frac{1}{2+\frac{1}{2+\frac{1}{2+\frac{1}{2}}}}=\frac{41}{29},$$

$$1+\frac{1}{2+\frac{1}{2+\frac{1}{2+\frac{1}{2+\frac{1}{2}}}}}=\frac{99}{70}.$$

These differ from $\sqrt{2}$ by

$$\left|\sqrt{2}-\frac{3}{2}\right| = 0.086 < \frac{1}{2^2} = 0.25, \quad \left|\sqrt{2}-\frac{7}{5}\right| = 0.014 < \frac{1}{5^2} = 0.040,$$
$$\left|\sqrt{2}-\frac{17}{12}\right| = 0.0025 < \frac{1}{12^2} = 0.0069, \quad \left|\sqrt{2}-\frac{41}{29}\right| = 0.00042 < \frac{1}{29^2} = 0.0012,$$
$$\left|\sqrt{2}-\frac{99}{70}\right| = 0.000072 < \frac{1}{70^2} = 0.00020.$$

(c) We have the continued fraction expansion (where $\gamma = (\sqrt{5}-1)/2$)

$$\gamma = \cfrac{1}{1+\cfrac{1}{1+\cfrac{1}{1+\cfrac{1}{1+\cfrac{1}{1+\cdots}}}}}$$

and continued fraction approximates

$$\frac{1}{1}, \quad \cfrac{1}{1+\cfrac{1}{1}} = \frac{1}{2}, \quad \cfrac{1}{1+\cfrac{1}{1+\cfrac{1}{1}}} = \frac{2}{3}, \quad \cfrac{1}{1+\cfrac{1}{1+\cfrac{1}{1+\cfrac{1}{1}}}} = \frac{3}{5}, \quad \cfrac{1}{1+\cfrac{1}{1+\cfrac{1}{1+\cfrac{1}{1+\cfrac{1}{1}}}}} = \frac{5}{8}.$$

These differ from γ by

$$\left|\gamma-\frac{1}{1}\right| = 0.38 < \frac{1}{1^2} = 1.00, \quad \left|\gamma-\frac{1}{2}\right| = 0.12 < \frac{1}{2^2} = 0.25, \quad \left|\gamma-\frac{2}{3}\right| = 0.049 < \frac{1}{3^2} = 0.111,$$
$$\left|\gamma-\frac{3}{5}\right| = 0.018 < \frac{1}{5^2} = 0.040, \quad \left|\gamma-\frac{5}{8}\right| = 0.0070 < \frac{1}{8^2} = 0.0156.$$

(d) We have the continued fraction expansion

$$e = 2 + \cfrac{1}{1+\cfrac{1}{2+\cfrac{1}{1+\cfrac{1}{1+\cfrac{1}{4+\cdots}}}}}$$

and continued fraction approximates

$$2+\frac{1}{1}=\frac{3}{1},\quad 2+\cfrac{1}{1+\cfrac{1}{2}}=\frac{8}{3},\quad 2+\cfrac{1}{1+\cfrac{1}{2+\cfrac{1}{1}}}=\frac{11}{4},\quad 2+\cfrac{1}{1+\cfrac{1}{2+\cfrac{1}{1+\cfrac{1}{1}}}}=\frac{19}{7},$$

$$2+\cfrac{1}{1+\cfrac{1}{2+\cfrac{1}{1+\cfrac{1}{1+\cfrac{1}{4}}}}}=\frac{87}{32}.$$

These differ from e by

$$\left|e-\frac{3}{1}\right|=0.28<\frac{1}{1^2}=1.00,\quad \left|e-\frac{8}{3}\right|=0.052<\frac{1}{3^2}=0.111,$$
$$\left|e-\frac{11}{4}\right|=0.0317<\frac{1}{4^2}=0.0625,\quad \left|e-\frac{19}{7}\right|=0.0040<\frac{1}{7^2}=0.0204,$$
$$\left|e-\frac{87}{32}\right|=0.00047<\frac{1}{32^2}=0.00098.$$

One sometimes sees the stronger inequality $\left|\text{number}-\frac{r}{s}\right|<\frac{1}{2s^2}$. This, however, is not generally valid. Take, for example, $\left|e-\frac{11}{4}\right|=0.0317$ and compare $\frac{1}{2\times 4^2}=0.0313$.

CPSIA information can be obtained
at www.ICGtesting.com
Printed in the USA
BVHW040528290119
538886BV00014B/220/P